UDL in the
CLOUD

UDL in the CLOUD

How to Design and Deliver Online Education Using Universal Design for Learning

Katie Novak and Tom Thibodeau

CAST Professional Publishing
UNTIL LEARNING HAS NO LIMITS™

Library of Congress Control Number: 2015957349
ISBN (Paperback) 978-0-9898674-8-1
ISBN (Ebook) 978-0-9898674-9-8

Published by:
CAST Professional Publishing
an imprint of CAST, Inc.
Wakefield, Massachusetts, USA

Photos of "Coco," "Kriti," "June," José," and "Ray" are licensed from iStockphoto.com.

Author photographs by Felix and Sara Photographers

For information about special discounts for bulk purchases, please contact publishing@cast.org or telephone 781-245-2212 or visit www.castpublishing.org

Cover and interior design by Happenstance Type-O-Rama.

Printed in the United States of America.

To Torin, Aylin, Brecan, Boden, Ariette, and Griffin.
Embrace the UDL philosophy. Know that if you come
to an obstacle in life, there is always a way to eliminate it.
Enjoy the journey, never give up, and remember that
mindset is everything.

Contents

Acknowledgments

Katie Novak

I am so lucky to be surrounded by such an amazing (and engaging!) support system. My mom is a fifth-grade teacher, who always gives me the inside scoop on how new initiatives in education affect the most important players in the game—our nation's teachers. She keeps my work honest. My coauthor, Tom Thibodeau (I also know him as *Dad*), works tirelessly to make education more accessible to students and faculty and I'm so grateful that he modeled UDL long before I had heard of it. I also have to thank David Gordon, the director of publishing and communications at CAST, as he is a wealth of knowledge, has a superpower for editing, and not only listens to, but encourages creativity and innovation in the field of UDL. Thanks to George Van Horn, a fellow member of CAST's UDL Cadre, for being the first to buy *UDL Now*, for introducing me to the concept of the virtual snow day, and for bringing the whole UDL team together every year for karaoke.

Lastly, my deepest gratitude goes to my husband Lon. Writing books is a daunting task when working full time and raising four kids, so this book is just another testament to how marrying him was the best decision I ever made.

Tom Thibodeau

I would like to acknowledge all the hard work that my wife Kathy puts into her fifth-grade class. It has truly been an inspiration to see how much she wants her students to succeed. I am also thankful to my coauthor and daughter Katie for her vast knowledge of UDL and her bottomless well of enthusiasm and energy. Dr. Don Vescio was my co-developer and co-teacher of the original *Teaching with Technology* course that we used in this book as the starting point of our UDL example, and I appreciate his collaboration. Thanks go to the editors and staff of CAST who have made the process of writing this book so enjoyable and educational. Lastly, to all the instructors, instructional designers, and students who I have learned with and from, thank you for your input. You have all helped me navigate my way through this process.

1

The Case for Better Online Course Design

IN THIS CHAPTER, we describe how education has changed through-out time and discuss some of the current issues with teaching online. Statistics and case studies are used to showcase the barriers that are very real and have implications for the future success of online learners. At the conclusion of the chapter, we profile five cloud learners to exemplify the significant variability in the online education community. We follow these learners throughout the text to exemplify how the Universal Design for Learning (UDL) framework may result in greater learning outcomes for all students.

"The Idea of what is true Merit, should also be often presented to [learners], explain'd and impress'd on their Minds, as consisting in an Inclination join'd with an Ability to serve Mankind, one's Country, Friends and Family; which Ability is (with the Blessing of God) to be acquir'd or greatly encreas'd by true Learning; and should indeed be the great Aim and End of all Learning."
— Benjamin Franklin, 1749

@PoorRichardUDL • Maintain our focus: true learning results when learners are inclined to increase their ability = succeed. #currenttranslation

Benjamin Franklin believed in the power of education, and he celebrated innovations that increased access to and availability of knowledge. In 1749, Franklin wrote *Proposals Relating to the Education of Youth in Pensilvania[sic]*, which garnered enough support to build an academy that eventually became the University of Pennsylvania, the Ivy League powerhouse we know today. In his proposal, Franklin penned, "The good Education of Youth has been esteemed by wise Men in all Ages, as the surest Foundation of the Happiness both of private Families and of Common-wealths."

Franklin had it right. Education has always been, and continues to be, a path to success and the bedrock of a free society. If the "pursuit of happiness" was a fundamental objective of the democracy Franklin helped create, then a robust and effective educational system was needed. Franklin and his fellow patriots believed that effective citizens and participants in society had to be informed, skilled, and able to articulate their opinions and demands. After all, he said, "The Constitution only gives people the right to pursue *happiness*. You have to catch it yourself."

In Franklin's day, providing a rich education to citizens was a daunting challenge, just as it is today. The population was spread out and communication was difficult. But the basic purposes of a sound and fundamental education have not changed. Today we talk about higher-level skills, critical thinking, complex problem-solving, collaboration, and communication as "21st century" skills, but when haven't they been essential?

As both a printer and a scientist, Franklin could appreciate the ways in which the technology of printed materials had expanded learning opportunities from the time of Gutenberg going forward. The large-scale manufacture of books and pamphlets accelerated learning and empowered average folks in ways that hadn't been possible before. We imagine Franklin would be thrilled—though not necessarily surprised—by what digital technologies afford us today. These technologies connect the world and put volumes of information literally at our fingertips, all at a relatively low cost. As the first Postmaster General of the United States,

Franklin would have loved email. If Poor Richard were alive today, such proverbs as *"Diligence is the mother of good luck"* and *"The learned fool writes his nonsense in better languages than the unlearned; but still 'tis nonsense,"* would be retweeted with alacrity.

@PoorRichardUDL · Diligence is the mother of good luck. #thetruth

MORE ENROLLMENTS, MORE VARIABILITY

We would like to think that Franklin, as one of the architects of a people's government, also would have been pleased with the demographic changes in today's students. In Franklin's time, education was reserved for a relatively homogenous population, one that was white, male, and well-heeled. Of course, that's no longer true. Today, we celebrate the diversity and variability of learners. We see individual differences and heterogeneous populations as a strength and an asset.

Nowhere are the opportunities and challenges offered by today's technologies more evident than in the field of online learning. Online enrollments are soaring. K–12 students are spending more and more time learning in the cloud. In 2000, there were only an estimated 50,000 K–12 students in virtual schools. By 2013, that number exceeded one million (Hawkins et al., 2013). This increase is related to the increasing opportunities that these students have to access education online. Cyber-charter schools, state-led virtual schools, and district-level supplemental online classes are now present in every state in America. In higher education, more than 21 million students took distance-learning courses in 2012–2013, most of them online (NCES, 2014).

As options grow to pursue online education, so does student variability (Archambault, Kennedy, & Bender, 2013). When virtual high schools originated nearly two decades ago, "virtual school students were described as highly motivated, honors/advanced, independent learners who were more likely to attend four-year college than their face-to-face counterparts" (Barbour, as cited in Hawkins et al., 2013, p. 64). Today, a much more diverse population enrolls in these programs. Students of color represent a larger proportion of online participants, as do students from less affluent socioeconomic backgrounds and those with documented disabilities (Molnar et al., 2013). Also, virtual schools are becoming available to younger and younger students. Recent research tells us that 26 states offer online schooling for students in grades K–5 (Hawkins et al., 2013).

There are also increasing trends toward hybrid-online (that is, blended) learning experiences in K–12. Chapter 7 will deal with these experiences exclusively, but we want to highlight them here briefly. Two hybrid-online offerings are the flipped classroom and the virtual snow day. In a "flipped classroom," traditional instruction is inverted. In traditional classrooms, students spend class time listening to lectures, which is a lower cognitive skill. The more difficult work, the application of that knowledge, is often done at home independently. When teachers flip their classrooms, students get their first exposure to new material when they are at home and then they can apply their knowledge with access to peers and the instructor, who can address misconceptions and provide mastery-oriented feedback. The use of flipping as a teaching model has "almost exclusively been tied to the incorporation of video or digital technology introduced prior to the in-class session" so students complete all initial learning activities in the cloud (Westermann, 2014, p. 44).

Districts that have virtual snow days require K–12 students to attend class online in inclement weather so they can continue learning despite the snow. Although these sessions are often asynchronous, elementary students are typically expected to complete five hours of work during the day, while their secondary counterparts are expected to complete six hours (Roscorla, 2014).

Whereas online courses and virtual high schools exist in learning management systems (LMSes), flipped classrooms and virtual snow days can take many different forms. Some districts use an LMS, such as My Big Campus or Canvas, whereas others use Google Classroom, Google Docs, or even Twitter through iPads and cell phones (Gumbrecht, 2015). It's important to note, however, that research has not yet confirmed the effectiveness of using social media in place of an LMS.

Given that learners of all ages are heading to the cloud for education, it's more important than ever that online instructors have a solid understanding of how to design learning opportunities to minimize barriers and maximize engagement.

PERFORMANCE GAPS IN ONLINE EDUCATION

The exponential growth of online learning begs a question: Do these courses result in the same learning outcomes as in face-to-face classrooms? After a review of research, it appears that the answer is no—but not because there are significant differences in the learning outcomes of students who *complete* online courses. The disparity is due to the fact that too many students do not persist and complete online courses. As noted in a recent review of literature (Jaggars, Edgecombe, & Stacey, 2013), nearly every study comparing course completion rates between online and face-to-face community college courses has concluded that online completion rates are substantially lower. Not only that, but these online courses affect student grades and their overall progress in their program of study. This phenomenon is not unique to online courses at the college level.

In a recent review of literature on the completion rates for K–12 virtual schools, Hawkins et al. (2013) write: "Although no official attrition statistics exist for virtual schools by state or school type, individual evaluations of some K–12 online learning programs indicate that attrition ranges broadly from 10% up to 70%" (p. 65). They cite two schools, the Illinois Virtual High School and the Alberta Distance Learning Center, which have 53% and 47% completion rates, respectively. Like the research completed at institutions of higher education, this research tells us that online education, as we currently know it, is failing many of our students. Something has to change.

Consider, too, the success of Massive Open Online Courses, or MOOCs. These highly produced online courses, some designed by the likes of Harvard, MIT, and Stanford using their most acclaimed faculty, draw huge numbers of students. Course rosters can swell to well over 40,000 students. Yet on average, the completion rates are dismal. In fact, a University of Pennsylvania study examined millions of users and found that only about 4% of students completed their courses (Lewin, 2013). Of the 155,000 students who signed up for an MIT course on electronic circuits, only 23,000 (15%) even finished the first problem set. Only 7,000 (5%) completed and passed the course (Carr, 2012). This may be, of course, because in the eyes of some, MOOCs are already passé because of their poor instructional design—that is, they lack interaction, which is a key element of good instruction.

Even when students persist in these courses, they are not always as successful as their peers in face-to-face courses. Although a meta-analysis of 20 years of research on distance learning suggests that students in distance-learning courses outperform their peers in face-to-face courses (Shachar & Neumann, 2010), there are other large-scale, peer-reviewed studies that suggest the opposite is true. For example, in an examination of 498,613 courses taken by 51,017 students in 34 community and technical colleges in Washington State, students were more successful in face-to-face classes than in online sections. When students persisted through to the end of the term (N = 469,287), the average grade was 2.95 (on a 4.0-point scale), with a gap between online courses (2.77) and face-to-face courses (2.97).

This gap, coined the *online performance gap*, was even more pronounced with male students, Black students, and students with lower levels of academic preparation, which exacerbated the performance gaps present in face-to-face courses (Xu & Jaggars, 2014).

Satisfaction outcomes also differed in a recent study. Keramidas (2012) found that online students were less satisfied with a variety of factors when taking online courses. These factors included instructor interaction, instructor enthusiasm, instructor approachability, quality of the program, and the evaluations of student performance. This perceived satisfaction rate may also be influenced by barriers that prevent some students from being successful in an online course: academic barriers, cultural barriers, financial barriers, technological barriers, instructional barriers, and institutional barriers (Irvin et.al, 2010). If one of the objectives of virtual learning is to increase accessibility and learning for non-privileged populations, while closing the learning gap, the current models have not been wholly successful (Xu & Jaggars, 2014). Some research, in fact, "[implies] that the continued expansion of online learning could strengthen, rather than ameliorate, educational inequity" (Xu & Jaggars, 2014, p. 651).

Because of the conflicting research outcomes of the online learning experience, it is clear that in at least some cases, the online performance gap is a disturbing reality. Let's take the fact to Twitter:

> Online courses, although they increase access, do not guarantee increased student success. #depressing

How's that for less than 140 characters? Now, we are in no way arguing that all online learning experiences are inferior, because research has found examples of students who were successful in online courses, but it's not all students and it's not in all learning environments. But we are asking why outcomes differ so wildly so we can also begin to explore ways to improve online learning.

Table 1-1 shows some of the barriers—academic, cultural, financial, technological, instructional, and institutional—that contribute to the problem.

TABLE 1-1 Types of Barriers in an Online Learning Environment

TYPE OF BARRIER	CONCRETE EXAMPLES
Academic barriers	Lack of student time, lack of executive function, lack of content preparation
Cultural barriers	Lack of understanding about the value of distance education, faculty's lack of readiness to communicate cultural sensitivity
Financial barriers	Inadequate interact access, availability of computers and necessary programs
Technological barriers	Infrastructure, equipment problems and maintenance, student computer skills
Instructional	Faculty training, faculty readiness to meet the needs of all learners, time commitment, technical expertise of faculty, design of course
Institutional	Availability of courses, cost of course, student support services

Source: Adapted from Irvin et.al (2010) and Bork & Rucks-Ahidiana (2013).

Given the barriers, can students succeed online? Throughout this book we argue that yes, they can. Or, as we might put it on Twitter:

@PoorRichardUDL · So many barriers can prevent students from succeeding in online learning environments. #timeforUDL

QUALITIES OF SUCCESSFUL ONLINE LEARNERS

Benjamin Franklin would have been a big fan of Facebook. He loved sending and receiving messages, and with modern technology, there wouldn't be weeks, months, or even years between replies. We also believe he would have been an avid "Googler," as his intellectual curiosity led him to investigate anything that piqued his interest. One such anecdote from Franklin's life weaves together his love of messaging and investigation; it involves his friend, Jan Ingenhousz, a Dutch chemist.

The story starts in 1783 with Ingenhousz's electrocution. After shocking himself, he did what most people do today after suffering a tragic accident—he posted it on Facebook. Well, of course there was no Facebook, but if it had been invented, we wager that he would have updated his status immediately. Since that wasn't an option, he set paper to pen to write to his friends about the experience.

Franklin, who already had an interest in medical electricity, was delighted to learn that Ingenhousz's acuity was increased after being shocked. Ingenhousz wrote: "It did seem to me I saw much clearer the difficulties of every thing, and what did formerly seem to me difficult to comprehend, was now become of an easy Solution" (Beaudreau & Finger, 2010). In today's world, Franklin could have pulled out his iPhone and responded via Twitter:

@PoorRichardUDL · The great @Ingenhousz just shocked himself into a hyper state of elation, awareness, and comprehension. #awesome

Instead, Franklin responded by letter and reminded Ingenhousz that he, Franklin, had once shocked himself intentionally to determine whether or not it was dangerous.

Together, they came to the same conclusion. Because electric shock did not appear to have negative consequences, and in Ingenhousz's case may have had positive consequences, the use of shock should be explored in clinical trials with "mad men" as a possible cure for melancholia (Beaudreau & Finger, 2010). To learn more about this, you will have to rely on Google, but we won't leave you hanging completely: Ingenhousz and Franklin discovered a new way to treat mentally ill patients—electroconvulsive therapy—that is still widely used today. This resulted because they were motivated, autonomous learners who were hungry for knowledge and wanted to explore learning in authentic, relevant, meaningful ways.

In order for online learners to be successful, they need to be equally motivated to pursue knowledge, ask questions, and connect with others in cyberspace to advance their learning. As instructors and course designers, we must find ways to engage and motivate online learners to help them succeed in this medium. Addressing the online performance gap that continues to create educational inequality for students who are unable to attend face-to-face courses is essential (Xu & Jaggars, 2014).

But what are the characteristics of a self-directed learner—besides a willingness to electrocute oneself in the name of science? And how can online instructors design courses that help all students internalize those characteristics?

We surveyed a number of articles about online learning to help us better understand the characteristics of successful online students and courses and to see if these relate to UDL. Many of these articles discuss the features of successful online course design without ever mentioning UDL. When viewed through a UDL lens, however, the recommendations provide some helpful guidance for shaping online education. Furthermore, our survey also points out the need for UDL.

Lawson (2005) suggests that coursework needs to be engaging and relevant—and that the structure of the course must support students so personal barriers do not impede course success. When students feel motivated and believe that courses are designed to integrate knowledge into real-life application, they express more competence and satisfaction with coursework.

Gruenbaum (2010) also finds that online students must be motivated, autonomous learners who are able to self-regulate their learning. Autonomous self-regulated learners seek help when they lack understanding, believe in their own abilities and have strong efficacy, know how to organize learning material and manage their time to complete assignments, and believe in the value of learning and education. These skills need to be explicitly taught and assessed for students to be successful (Gruenbaum, 2010).

Roper (2007) identifies seven characteristics of successful online students compiled from survey data collected from 59 successful college graduates who took at least 80% of their coursework online. Successful students noted the importance of managing time, participating in discussions, applying knowledge, asking questions, staying motivated, and making connections with the instructor and other students. (Of course, by surveying only those students who graduated, the authors lose the perspective of those students who were not successful.)

If students persist because they are motivated, autonomous, and self-regulated learners, one can assume that those students who drop out of online courses lack those characteristics and don't have instructors who can help them learn such important internal functions. At a large Midwestern university with online dropout rates of 54.2%, Park and Choi (2009) examined whether persistent learners and dropouts had different individual characteristics, external factors, and internal factors (i.e., motivation). Results suggest that many different barriers prevent students from persisting in an online course. In addition to motivation, relevance and organizational support were the strongest predictors of student persistence.

Course developers and instructors can influence both of these factors. Faculty involved in the development and delivery of online coursework need to ensure that content is meaningful, authentic, and relevant and also that the structure of the course eliminates barriers and does not allow for student failure. That's where this book can help.

HOW UDL CAN HELP

By employing Universal Design for Learning (UDL) principles and guidelines, we can better reach online learners who are "in the margins." And just to make it clear: although the UDL Guidelines are widely implemented in K–12 settings, the biggest area of growth in using UDL is in the postsecondary world. In fact, the UDL Guidelines were developed by David Rose, a longtime lecturer at Harvard Graduate School of Education, in his courses. He notes that the Guidelines are deeply influenced by what he learned in a postsecondary setting and are just as applicable to adult learners. This book examines the variability of our new online learners, discusses the environments in which they are currently learning, and provides concrete strategies to help educators meet the needs of this growing cohort of online learners.

UDL is a framework that allows all learners to access rigorous material. This framework is built upon a list of strategies, or Guidelines, that help to make curriculum and instruction accessible to all students, regardless of variability. When instructors implement these Guidelines, student outcomes increase.

So, it's a no-brainer, right? All that instructors need to do is implement a list of strategies and online courses become a factory of student success. Oh, if only it were that easy.

First, the UDL Guidelines require changes to instructional approaches. Many instructors deliver courses using traditional methods of teaching and learning. Take for instance the term *lecturer*. Just the one word paints a picture of someone standing in front of a room, delivering a monologue, sharing knowledge. There are numerous drawbacks to this technique, yet it is still widely used.

The UDL Guidelines also require a philosophical shift in the way we view teaching and learning. Online educators must believe that it is their responsibility to design and deliver courses that eliminate the barriers that students bring to the learning environment. Educators must embrace the belief that they are not only content experts, but learning experts, who are responsible for imparting knowledge and setting up an environment in which all students can learn this knowledge and express it back in a meaningful, relevant way.

INTRODUCING OUR "STUDENTS"

The UDL principles and Guidelines will be discussed in more detail in subsequent chapters. But first, we'd like to introduce some students whose experiences can help us better understand the challenges of learning online—and how good course design can support their education. Though they are not actual people, their profiles and experiences mirror those of real students we have known and worked with. We will follow these students throughout the book to illustrate challenges and strategies of teaching and reaching all learners.

Coco and Kriti are K–12 students. June, José, and Ray are adult students taking the course "Teaching with Technology" at the hypothetical Poor Richard's University. We introduce them here and revisit them in subsequent chapters, using their stories to help us better understand virtual course design from a UDL perspective.

STUDENT	BARRIERS
	Meet Coco. She is a fourth grade-student outside of Boston. Coco is a perfectionist and is very anxious about her grades. She often cries at night when completing her homework because she is convinced the teacher will say it's wrong. Her school district has recently invested in a learning management system (LMS) so students can engage in "virtual snow days" so they don't have to make up days when the weather prevents schools from opening. Coco is very nervous about the newly adopted "virtual snow days" because she has only one computer at home, and she has to share it with her two brothers. Because she's the youngest in her family, she is usually last in line for the computer, and she knows she won't be able to spend the whole day on the computer.

STUDENT	BARRIERS
	Meet Kriti. She is 17 and wants to someday be an early childhood teacher. Kriti and school have never really gotten along well. She survived through her junior year, but just barely. Kriti's ADHD is under control, but the medicine makes her feel like the world is running in slow motion most of the time. She falls asleep easily in class if things are not active and has a bad habit of procrastinating on assignments. Because she has had difficulty in school, she has decided to finish her diploma in a virtual high school. Her parents are not supportive because they feel like the virtual classes will be "a joke."
	Meet June. She is 62 years old and has been teaching English composition at her local community college for 25 years. June and computers are relative strangers who meet each day for coffee so June can do her required administrative activities, attendance, and lesson management. Because her computer skills are weak, she still prefers to correct her student assignments on paper with a blue pencil. She worked as an assignment editor at a local newspaper for 15 years before she began teaching. She loves the printed word. Her college has just decided that all courses must now have an online presence in the college's LMS. The training sessions are being held during her in-class sessions so she will have to take a course online.
	Meet José. José is 34, single, well-adjusted, and the life of the party. He had a great job as a butcher, where he was in line to become the department manager before the supermarket closed down two months ago. José is a hard worker who did well in high school and college. He loves to read and loves art. When he got out of school, he wasn't able to use his degree in graphic design in a down economy so he followed his dad into the family business. He has a second chance with a state labor retraining grant and thinks that being an instructional designer would be a great opportunity to combine his past degree with a new one.

STUDENT	BARRIERS
	Meet Ray. Ray just got out of the Army. He didn't know what to do after high school so he thought the Army would help give him some focus. He is married with two boys, eight and six. He did two tours in Afghanistan and was wounded on the last tour. He worked as an assistant to a major and was wounded en route to a new camp. The bullet wound in his leg has healed well, but he will get around slowly for at least another year. He came home with a Purple Heart and is eligible for the the G.I. Bill, but has no real plans for the future. He loves computers, though. He has a new job in tech services at a local school district, but without a college degree, he will be fixing hardware and software issues forever. He can't afford not to work, so his schedule will not allow him to take face-to-face courses. Because he wasn't the best student in high school and does not enjoy reading, he is nervous about going back to school, especially online, and fears that he will not be able to complete the courses due to lack of time.

Is there a way to meet the needs of all these students online? Yes, there is. You just have to follow the wise words of Benjamin Franklin: "An investment in knowledge pays the best interest." Invest your time in reading this text and learning about UDL to help all your learners have the best educational outcomes possible. And if you're on Twitter, you may want to share the following:

UDL is for online learners too? I'm in. #readytochangetheworld

The question for us is how to use the UDL Guidelines to implement the features of a successful online course and scaffold the executive functions that students need to be successful. If this is done correctly, barriers will be eliminated and the online performance gap will wane, while our students will gain important skills so they can compete in the future.

If we return to the profiles of our students, it is clear that many may lack motivation, autonomy, and self-regulation strategies, while also facing additional barriers. To help all students succeed, the courses must be designed to minimize or eliminate these barriers.

STUDENT	BARRIERS
	Coco's anxiety prevents her from being an autonomous learner. In class, she asks numerous questions and spends a lot of time at the teacher's desk to make sure she understands what is expected from her. Also, since there is only one family computer at home that she has to share with her brothers, there will be a technological barrier, because not all three children will be able to be online for the required 5–6 hours. If she is able to be online, she will expect constant reassurance.
	Kriti lacks motivation, autonomy, and self-regulation strategies. In addition to academic barriers, Kriti has to overcome cultural barriers, since she doesn't truly see the value of online education and her family doesn't see the value either. Her instructor will have a lot to overcome to engage Kriti in this environment.
	June faces technological barriers because of inadequate technology skills. She is motivated and autonomous, but may lack skills on how to regulate her learning online without clear scaffolding from the instructor.
	José is motivated, autonomous, and can self-regulate his learning. The challenge with José is how to keep him engaged and motivated in a class with students who face multiple barriers and need support and scaffolding in areas where José is already successful. How can our instructor meet José's needs while also meeting the needs of the other students?

STUDENT	BARRIERS
	Ray works full time, so he lacks the time to return to school to pursue a degree so he can move ahead in the field. He wasn't the most successful student in high school, although he is motivated. Also, he needs to learn self-regulation strategies to help him persist in the course while balancing full-time work and family.

In the next chapter, we'll discuss how UDL helps us design courses or online lessons that align with research on best practices so that all of our students can become self-regulated, motivated learners. Our greatest hope would then be that we teach our students content, but that we also teach them how to be motivated learners so they can continue to learn long after the course is over.

As the authors of the UDL Guidelines noted: "The goal of education in the 21st century is not simply the mastery of content knowledge or use of new technologies. It is the mastery of the learning process. Education should help turn novice learners into expert learners—individuals who want to learn, who know how to learn strategically, and who, in their own highly individual and flexible ways, are well prepared for a lifetime of learning" (CAST, 2011).

Learners should get engaged—though not necessarily electrocuted—in their pursuit of education. Well-prepared educators can help them achieve this.

REFLECTION

- To be successful in an online course, students have to be motivated, autonomous learners. Those who are not require explicit scaffolding to acquire these skills. We argue it is the responsibility of the instructor to design an online environment in which all students can learn how to be learners. To do this, you must be both a content expert and a learning expert. Reflect on this statement and assess your ability to teach students how to learn. What are your strengths? What areas will you need to focus on to increase student motivation and autonomy?

OPTIONS FOR EXPRESSION

- To eliminate barriers in your learning environment, you need to have a growth mindset and believe that you are capable of designing a course where all students will be successful. Explore the work of Carol Dweck and read her book, *Mindset: The New Psychology of Success.* Developing a growth mindset will be valuable as you deliver universally designed online instruction.

- Table 1-1 presents common barriers to online education. Review each barrier and consider how you can begin the work of eliminating one or more of those barriers at the course or institutional level. As a first step, draft a letter, create a presentation, or host a discussion with colleagues to increase the number of stakeholders who can collaborate to eliminate the barriers on a systematic level.

- Collect and analyze data on completion rates, student outcomes, and grades in your online course or in all the online courses in your institution. Compare the data to the completion rates and student outcomes in face-to-face courses. Share the data with colleagues in the form of a multimedia presentation or memo.

Fewer Barriers, More Support: UDL Guidelines in Action

2

MANY CLOUD INSTRUCTORS do not have a background in distance education theory and practice. These instructors are content-area or grade-level specialists who are more adept at delivering face-to-face instruction. In this chapter, we explore online discussion strategies to help instructors realize how many of these strategies create significant barriers for our learners. We discuss the UDL Guidelines and review research that exemplifies how implementing the Guidelines will help deliver content and improve student outcomes.

In Chapter 1, we highlighted both the opportunities and challenges facing online educators. It is not enough to point out the need, however. Educators need professional training and development in how to build and deliver online courses that meet the needs of all learners and eliminate all barriers. We hope you'll find that this book is a great first step in that direction.

Although research suggests that course design requires a scientific approach, there is "scant evidence in the field of curriculum development and evaluation of a scientific imprint" (Doabler et al., 2015, p. 98). In order to successfully design courses for all students, educators must understand the science of curriculum design. Yet many instructors lack explicit instruction on the curriculum design process.

As a result, online coursework tends to be designed to work best for students who are self-directed learners, who are tech savvy, and who have other skills necessary to be successful in the cloud (Keramidas, 2012). In the same way that traditional curriculum is often designed to serve "average" students—leaving the concerns of those with exceptional abilities and needs out of the planning loop—online learning course design often fails to consider the dramatic variability of learners and the special circumstances of the online setting itself.

UDL provides a scientific framework for course design that will increase access, participation, and success for all learners. In fact, the Higher Education Opportunity Act of 2008 (HEOA) both defines UDL and endorses UDL implementation for pre-service preparation of teachers, in-service teacher training, and postsecondary instruction. The act endorses not only teaching *about* UDL but teaching in a UDL way. The HEOA defines UDL as follows:

> Universal Design for Learning (UDL) means a scientifically valid framework for guiding educational practice that:
>
> (A) provides flexibility in the ways information is presented, in the ways students respond or demonstrate knowledge and skills, and in the ways students are engaged; and
>
> (B) reduces barriers in instruction, provides appropriate accommodations, supports, and challenges, and maintains high achievement expectations for all students, including students with disabilities and students who are limited English proficient.

K–12 policy documents, such as the National Education Technology Plan (2010) and drafts of the reauthorization of the Elementary and Secondary Education Act (not yet passed as we went to print), also strongly endorse UDL.

@PoorRichardUDL · Reduce the barriers and succeed with UDL. #flexibleUDL

THE UDL PRINCIPLES

Universal Design for Learning can be summed up in the old adage often attributed to Ben Franklin (but more likely from an ancient Chinese philosopher): "Tell me and I forget. Teach me and I remember. Involve me and I learn."

This statement does not speak specifically to teaching children or teaching adults but rather adopts a universal first person voice. Any learner can speak to this truth, "Involve me and I learn." There are colliding theories suggesting that children and adults learn differently: pedagogy and andragogy. *Andra* translates as the word adult, which makes *andragogy* the art and science of teaching adults, whereas *peda* translates as child, which makes *pedagogy* the art and science of teaching children (Knowles, 1980). Taken more broadly, andragogy is sometimes referred to as learner-focused education, whereas pedagogy is referred to as teacher-focused education (Taylor & Kroth, 2009).

UDL doesn't make such distinctions. From the UDL perspective, all learning experiences are learner-centered. (For more on the andragogy vs. pedagogy debate, see Appendix.)

When educators design instruction using UDL, they involve all learners, regardless of age. The UDL framework assumes variability in motivation, interest, and readiness, and so provides scaffolding in language function, explicit skills, and executive function while keeping learner engagement and choice at the forefront. If, as educators, we hold the highest of expectations for all learners, both K–12 students and adults, we can engage, motivate, and ultimately facilitate authentic, relevant learning experiences for all learners.

Benjamin Franklin was an early supporter of education for both young students and adults because he believed that education had the potential to solve the nation's problems (Blinderman, 1976). Although UDL would not be established as a framework for centuries, the education Franklin envisioned and facilitated aligns with many of today's UDL Guidelines. These Guidelines will be discussed in detail throughout the chapter; but first, let's take a moment to view these Guidelines (Figure 2-1).

Universal Design for Learning Guidelines

I. Provide Multiple Means of **Representation**	II. Provide Multiple Means of **Action and Expression**	III. Provide Multiple Means of **Engagement**
1: Provide options for perception 1.1 Offer ways of customizing the display of information 1.2 Offer alternatives for auditory information 1.3 Offer alternatives for visual information	4: Provide options for physical action 4.1 Vary the methods for response and navigation 4.2 Optimize access to tools and assistive technologies	7: Provide options for recruiting interest 7.1 Optimize individual choice and autonomy 7.2 Optimize relevance, value, and authenticity 7.3 Minimize threats and distractions
2: Provide options for language, mathematical expression, and symbols 2.1 Clarify vocabulary and symbols 2.2 Clarify syntax and structure 2.3 Support decoding of text, mathematical notation, and symbols 2.4 Promote understanding across languages 2.5 Illustrate through multiple media	5: Provide options for expression and communication 5.1 Use multiple media for communication 5.2 Use multiple tools for construction and composition 5.3 Build fluencies with graduated levels of support for practice and performance	8: Provide options for sustaining effort and persistence 8.1 Heighten salience of goals and objectives 8.2 Vary demands and resources to optimize challenge 8.3 Foster collaboration and community 8.4 Increase mastery-oriented feedback
3: Provide options for comprehension 3.1 Activate or supply background knowledge 3.2 Highlight patterns, critical features, big ideas, and relationships 3.3 Guide information processing, visualization, and manipulation 3.4 Maximize transfer and generalization	6: Provide options for executive functions 6.1 Guide appropriate goal-setting 6.2 Support planning and strategy development 6.3 Facilitate managing information and resources 6.4 Enhance capacity for monitoring progress	9: Provide options for self-regulation 9.1 Promote expectations and beliefs that optimize motivation 9.2 Facilitate personal coping skills and strategies 9.3 Develop self-assessment and reflection
Resourceful, knowledgeable learners	**Strategic, goal-directed learners**	**Purposeful, motivated learners**

FIGURE 2-1 The UDL Guidelines. © 2013, CAST. Used with permission.

The Guidelines are organized into three distinct principles, which are defined by columns:

- Provide multiple means of engagement (how will they will interact or persist in the course or lesson?),

- Provide multiple means of representation (how will they deal with the content of the course or lesson?), and

- Provide multiple means of action and expression (how will they interact with the content of the course or lesson?).

These principles align to learning "networks" identified in the cognitive sciences: the affective network, the recognition network, and the strategic network.

If you want students to learn, all three networks need to be activated. You activate the three networks by implementing the UDL principles. See the relationship between the networks, the principles, the Guidelines, and the checkpoints in Table 2-1.

TABLE 2-1 Relationship between UDL Principles and Brain Networks

TO ACTIVATE THE BRAIN: ADHERE TO UDL PRINCIPLES →	TO ADHERE TO UDL PRINCIPLES: IMPLEMENT GUIDELINES →	TO IMPLEMENT THE GUIDELINES: IMPLEMENT THE FOLLOWING CHECKPOINTS
Provide multiple means of engagement	Provide options for self-regulation	• Promote expectations and beliefs that optimize motivation • Facilitate personal coping skills and strategies • Develop self-assessment and reflection
	Provide options for sustaining effort and persistence	• Heighten salience of goals and objectives • Vary demands and resources to optimize challenge • Foster collaboration and community • Increase mastery-oriented feedback

TO ACTIVATE THE BRAIN: ADHERE TO UDL PRINCIPLES →	TO ADHERE TO UDL PRINCIPLES: IMPLEMENT GUIDELINES →	TO IMPLEMENT THE GUIDELINES: IMPLEMENT THE FOLLOWING CHECKPOINTS
	Provide options for recruiting interest	• Optimize individual choice and autonomy • Optimize relevance, value, and authenticity • Minimize threats and distractions
Provide multiple means of representation	Provide options for comprehension	• Activate or supply background knowledge • Highlight patterns, critical features, big ideas, and relationships • Guide information processing, visualization, and manipulation • Maximize transfer and generalization
	Provide options for language, mathematical expressions, and symbols	• Clarify vocabulary and symbols • Clarify syntax and structure • Support decoding of text, mathematical notation, and symbols • Promote understanding across languages • Illustrate through multiple media
	Provide options for perception	• Offer ways of customizing the display of information • Offer alternatives for auditory information • Offer alternatives for visual information
Provide multiple means of action and expression	Provide options for executive functions	• Guide appropriate goal-setting • Support planning and strategy development • Enhance capacity for monitoring progress

TABLE 2-1 Relationship between UDL Principles and Brain Networks   *CONTINUED*

TO ACTIVATE THE BRAIN: ADHERE TO UDL PRINCIPLES →	TO ADHERE TO UDL PRINCIPLES: IMPLEMENT GUIDELINES →	TO IMPLEMENT THE GUIDELINES: IMPLEMENT THE FOLLOWING CHECKPOINTS
	Provide options for expression and communication	• Use multiple media for communication • Use multiple tools for construction and composition • Build fluencies with graduated levels to support practice and performance
	Provide options for physical action	• Vary the methods for response and navigation • Optimize access to tools and assistive technologies

Think of each network as a game of high striker. You may remember this game from carnivals, where you were given a hammer or mallet and you had to hit a target to try to ring the bell on top. In order to "ring the bell" to activate the correct part of the brain, educators need to take a proverbial mallet and hit the corresponding UDL principle in that column. To do that, you must implement the UDL Guidelines.

@PoorRichardUDL · Play UDL high striker for a strong mind. #udlpuns

Simply examining the UDL Guidelines will not result in a universally designed course for learners. Translating these Guidelines into effective design practices requires training, time, and perhaps new technical development. Edyburn (2010) makes a poignant observation regarding the supposition that UDL is just good teaching: "Since UDL is the convergence of multiple disciplines, I reject the notion that there is a natural trait within effective teachers that allows them to implement UDL without knowing that they are doing so" (p. 38).

So, where do we start to ensure that readers of this text learn UDL theory, practice, and its implications for course development? This book offers awareness training about the importance of UDL in the cloud and provides a framework for the technical development of online learning experiences to meet the needs of all learners. As for time, it is our hope that when instructors become aware of the UDL principles and how to implement them, they will set aside the time to redesign courses to make them more universally accessible and effective.

@PoorRichardUDL • UDL does take time to implement but the time is always manageable. #todoUDLstartUDL

UDL APPLICATION: THE EXAMPLE OF ONLINE DISCUSSION BOARDS

We find that the best way to begin applying UDL principles and guidelines is to look at a typical "learning experience" for online learners and note the barriers that are present. From there, we can discuss how to eliminate the barriers using the UDL Guidelines and to intentionally design something better: a supportive learning environment that builds on learners' strengths and differences. When this practice is completed with all aspects of teaching and learning, all students are able to access rigorous curriculum.

To show how this works, let's consider the online discussion board. The purpose of discussion boards is to provide a way for students to interact and discuss components of the course since they do not meet in person. When designed correctly, these discussion postings can increase student achievement. However, this will not occur if students are only "surface posting," or posting only because it is a requirement (Blackmon, 2012).

Often, in online sessions, whether it's a full course or a virtual snow day, the explanation of the discussion board focuses on requirements, which would likely result in these surface postings.

In an examination of discussion board instructions from an online course, the following information was provided to students regarding their participation.

• Complete all readings and post your initial thoughts, questions, or comments on the discussion board by the due date. Must be a minimum of

200 words. Postings less than 200 words will not receive more than 3/6. Postings less than 100 words will receive a 0.

- Post *at least two* thoughtful follow-up comments to the discussion board by the due date. Both comments must be at least one paragraph in length. Aim for a minimum of 100–150 words in each response. Responses less than 100 words will receive a 0.

The instructions above may look eerily familiar to you, as they are typical instructions for online discussions that reflect deadlines and minimum word counts. Let's examine some possible barriers to these instructions, however, in terms of the UDL Guidelines. Specifically, let's just look at engagement (Table 2-2) and whether or not the instructions provide options for recruiting interest, sustaining effort and persistence, and self-regulation. To do this, let's ask ourselves the questions in Table 2-2.

TABLE 2-2 Check Your Implementation: Engagement

ASK YOURSELF THE FOLLOWING QUESTIONS TO DETERMINE WHETHER UDL PRINCIPLES WERE IMPLEMENTED.	
Did the instructions provide multiple means of engagement?	**Provide options for recruiting interest** • Did students have a choice about how they would respond? Podcast? Typed response? Audio? **Provide options for sustaining effort and persistence** • Were the goals of the discussion board clear? (In other words, what is the point of online discussions in general?) • Did the instructions vary in terms of challenge? • Did the instructor note he/she would assess and respond with mastery-oriented feedback? **Provide options for self-regulation** • Was there a rubric so students could reflect and assess their work before they posted it? • Did the instructor provide troubleshooting to support students if they could not log in or access the discussion board?

After reading the questions in Table 2-2, the answers are resoundingly similar: *No*. The instructions for the discussion board were not universally designed to increase engagement. When examining the research on what makes discussion board contributions more successful, however, findings align to the questions we asked above.

Findings of a recent study indicate that discussion postings result in increases in complex thinking, substance, and quality when instructors actively respond to posts and provide feedback and when grading rubrics are used to measure the quality of postings (Giacumo, Savenye, & Smith, 2013). These findings relate directly to two Guidelines-based questions posed above:

- Did the instructor note he/she would assess and respond with mastery-oriented feedback?

- Was there a rubric so students could reflect and assess their work before they posted it?

These two questions are just examples of barriers you might identify right from the beginning that may affect student engagement, and therefore, student achievement.

So, how do we respond to these barriers once we're aware of them?

First, we need to communicate to students that as instructors, we will respond to posts with feedback to guide their thinking to be more complex and thoughtful. This feedback can be provided both during the discussion and at the end of the discussion period as appropriate to the class and the discussion assignment.

Second, we can provide a rubric that does not just outline surface requirements, but requirements for quality and content. Again, these two shifts in practice are intentional because we realize that without them, all students may not be engaged in the discussion board, and therefore, their engagement in the course would decrease.

See the following addition, which could precede the original instructions. This could be posted via text, and could also be linked to a video and/or audio component so multiple means of representation are provided.

- A portion of your grade will be based on your participation in an online discussion board. This is a place where you can post your thoughts about the week's content and respond to classmates by typing a response, posting a podcast, recording audio, creating a PowerPoint, and so on. Whatever medium you choose, I encourage questioning, academic discourse, and replies that push us to think deeper, as long as postings are respectful and non-discriminatory. I will read each of these postings and responses and provide feedback so you can think deeper about complex topics in the course.

- For the Week 1 discussion, you will be introducing yourself to your classmates, so we can get to know each other, and so you have prac-tice with the tool before it counts toward your official grade. As you are drafting this response, view the discussion board rubric and the provided exemplar, so you understand what is expected in future discussion postings.

Following this introduction, students could view a rubric and an exemplar that help to scaffold the process of writing a quality discussion and responding to the original posting in a collaborative way.

Once this information is presented, it would be important to revise the original instructions so they don't contain so many barriers, or rigid expectations for postings.

- If you would prefer to hear a short podcast of these requirements, click *here*.

- Complete all readings and post your initial thoughts, questions, or comments on the discussion board by the due date. Generally, these postings should be complex and should reflect depth of thought. This cannot be accomplished in a few words. If you are typing a response, try to write at least a page. If you choose to post video or audio, or use another medium, try to imagine you're participating in class. Your "posting" should last for a couple of minutes. Please see the rubric for specifics on how the complexity of postings will affect your grade.

- Post *at least two* thoughtful follow-up comments to the posts of your classmates to the discussion board by the due date. Again, you can post these comments by typing in the discussion board or you can use another medium such as a podcast or video response. Again, for specifics on how these replies will be graded, view the discussion board rubric.

Discussion postings are just one small piece of the online learning puzzle, and yet, there are so many aspects that could be designed for improved student access. In the previous example, we only focused on increasing student engagement in discussion boards by modifying the instructions. If we continued to peel back the layers of barriers that may prevent students from learning, we would see the need for UDL more and more.

@PoorRichardUDL · Like a peeled onion the sweetest part of learning is at the core of UDL. #UDLpracticemakeslearningbetter

SPECIAL ASSIGNMENT: INCREASE ENGAGEMENT

Implementing UDL does not happen by accident. Increasing engagement must happen if all students are to persist in the discussion board environment. Benjamin Franklin seemed to understand this centuries ago. In 1723, Franklin wanted to test his beliefs about education by organizing discussion groups where students were encouraged to engage in scientific and philosophical discussions. Granted, these discussions were in person, but with advances in technology, online discussions of today can offer the same opportunities for engagement.

First, Franklin offered multiple means of engagement by **fostering collaboration and community** and **minimizing threats and distractions** by proposing questions to all interested candidates such as, "Do you sincerely declare that you love mankind in general, of what profession or religion whatsoever?" and "Do you love truth for truth's sake and will you endeavor impartially to find and receive it yourself, and communicate it to others?" (Blinderman, 1976, p. 15.) This helped to bring everyone toward a common mission where they would be encouraged to speak the truth without fear of retaliation, and that they would be surrounded by members who valued people regardless of profession or religion. Franklin also **recruited participant interest** by setting aside time for eating, drinking, and being merry at each discussion to increase the value of the discussions to the participants.

Now, we can't offer snacks and libations to our online students, but we can recruit participant interest, foster collaboration and community, and minimize threats and distractions all within the instructions for the discussion postings and in many other online environments. We can also redesign the course to provide multiple means of representation, action, and expression. When done effectively and deliberately, the course would implement the UDL Guidelines and students would have equal opportunities to succeed. This process of course redesign will be discussed beginning in Chapter 3.

To end this chapter, let's reflect back on a few of our case studies from Chapter 1 to examine how these small shifts in building engagement, using the UDL Guidelines, would make an online discussion more accessible for our learners.

Coco is required to attend virtual snow days at her school.	Coco's teacher wants her students to review assigned material and contribute to the discussion board. This will be frustrating for Coco because her keyboarding skills are weak, and since she will have to share the computer with her siblings, she won't have the necessary time to meet the day's requirements in the LMS. This increases her anxiety about the possibility of a storm. If she could write the response, however, and be allowed to turn it in the next day, or record her response on a phone and upload it at night, she wouldn't worry so much about failing.

June, you will remember, is our 62-year-old adjunct instructor who teaches English composition.	To reach June, we need to provide her with a warm and inviting online course that is easy to look at, easy to follow, and has lots of options. She will need tutorials and support. Introductions need to present the activities of the week and be available as videos and text for her to watch, print out, or both and read again at her leisure. She will need to be invited into the discussion boards with several guided "test" topics in the first week to help her get acclimated to the process. The technology barrier may need to be addressed personally with June in a one-on-one phone call that walks her through the process or a face-to-face meeting. Once she is accustomed to the discussion boards she will quickly realize how much she likes to communicate this way and it will become her favorite part of the class.
Ray is our army vet. He is a hard worker and loves computers, but he's nervous about succeeding because of his family and work obligations.	To help Ray, an online course will have to not only define all of the steps necessary to do the assignments but it will have to challenge him and motivate him to find the time between work and family to do all the coursework. He will particularly need multiple forms of representation to engage him in the content. Since he is not a reader or a writer, the course will have to provide options for audio, visual, and kinesthetic modes of representation and expression. Getting Ray to participate in the discussion boards will be a challenge for the instructor, but if the questions that start the discussion board are open-ended and interesting, the environment is welcoming and non-threatening, and the participation expectations are concise and clear, Ray can be persuaded to participate and succeed in the process. Giving him the option of submitting a podcast or audio file will also go a long way to making Ray feel at home in the discussion board.

These students represent the wide variety of students we will see in online learning environments. They come from every walk of life and have dreams of success across the entire range of possibilities. However, they all come with either a perceived or real barrier to that success. It is our job, using the UDL Guidelines, to find the right combination of options to reach them *simultaneously*!

@PoorRichardUDL • UDL can manage all the learning needs of your students. #UDLGuidelineshelp

REFLECTION

- Which Guidelines would be the most difficult for you to implement in a cloud environment and why?

- The discussion board is one example of a learning environment that may contain barriers that prevent all students from learning in a virtual environment. After reading this chapter, identify common barriers in discussion boards. How can the UDL Guidelines support educators as they redesign expectations so all learners can access the discussion and move beyond surface postings?

OPTIONS FOR EXPRESSION

- To engage students, you will need to know what is relevant and meaningful to them. Create an online survey or Google form that asks students to share their goals for taking your course, how they plan to manage their time, and how they hope to apply the knowledge gained in the course in the real world. Distribute the tool and examine the answers for themes to help you provide choices and scaffolds throughout the course to meet the needs of all learners.

- To learn more about the UDL Guidelines, visit www.udlguidelines. org. On this site, you can find examples and resources that illustrate each of the UDL checkpoints and read the latest evidence and scholarly research.

3

How to Develop a Syllabus the UDL Way

THIS CHAPTER WILL offer a practical protocol for developing a course syllabus using the principles of UDL. The focus will be on objectives and process before content and course construction. If you are preparing for virtual snow days or a flipped class in an elementary or secondary setting, you may want to think of this process as a detailed lesson-plan construction.

If we want all of our students to love knowledge so much that they are willing to electrocute themselves in the name of learning, like our friends Franklin and Ingenhousz, we need to work some serious magic when designing our courses. When designing an online learning experience, we can't just focus on how students will access the course. Instead, we must go into the design process by thinking about how we can ensure that all students persist with the course and are successful in meeting the course objectives. We need *all* students to be able to succeed—not just those who are already self-regulated, autonomous learners. Those students will learn regardless of the quality of the instructor, but great instructors can inspire and motivate all students, regardless of variability.

Moving from access to success requires shifts in the way online courses are designed and delivered. This chapter will focus on building online learning models based on the research presented in the previous chapters. Designing an online course or even a snow-day lesson using the UDL framework requires instructors to heighten the salience of goals and objectives when considering course design.

> **@PoorRichardUDL** • Be at war with your vices, at peace with your neighbors, and let every new year find better course design. #keepontrying

A great place to begin is the building of a course syllabus. Whether instructors teach virtual high school courses, college or graduate courses online, or simply an online session as a virtual snow day, the same principles apply. Building an online course is not just about presenting content. It's also about scaffolding self-directed learning skills so all students can access that content.

Xu and Jaggars (2014) discuss the importance of embedding scaffolding into course design to close the online achievement gap. In order for scaffolding to be effective, however, online faculty must understand how to explicitly model and support students as they develop their courses. This chapter will provide concrete suggestions for building this scaffolding into a course from the beginning.

As we mentioned in previous chapters, to be successful in online courses, students must have self-directed learning skills, and yet, a recent study found that many online instructors do not believe it's their responsibility to teach these skills (Bork & Rucks-Ahidiana, 2013). If this belief is widespread, students who do not possess the executive function to succeed in a self-directed environment do not have a chance. Their lack of self-direction will act as a barrier, and will prevent them from succeeding in an environment that is designed, in theory, to increase their access to education. The other reality is that many current faculty may not know how to teach the self-directed skills necessary for online success simply because they were never taught these skills explicitly in their own course work. Obviously, the primary reason why this may be true is that many of the current online faculty never took an online course! Additionally, many faculty, especially at the college and graduate level, have usually not been taught how to teach either. They have been trained as content experts or researchers (Vella, 1990).

Ideally, all education institutions should provide proper support for instructors in the development of their courses, but many do not. This creates a barrier for the instructors who are tasked with teaching students how to learn when they themselves have not been taught to teach or learn in online environments. This creates a gap between the expansion of online education and the ability of institutions to meet the needs of both students and faculty who teach these courses (Orr, Williams, & Pennington, 2009).

This support is not something that can be implemented overnight. Faculty need ongoing support from a properly focused instructional designer over time. They need time to:

- Develop a working relationship between the instructor and instructional designer;

- Complete a literature review on current best practices in online education;

- Understand the intended educational outcomes of the course;

- Create learning models and develop the course;

- Rethink the needs of the course in an online environment;

- Reenvision the role of the instructor in the online course;

- Reconstruct the course activities to be effective online; and

- Evaluate the learning models in response to student outcomes (Vasser, 2010).

That being said, this is time well spent because the greatest potential for institutional improvement is strategic, ongoing communication concerning online education (Orr, Williams, & Pennington, 2009). It is our hope that this text can prompt those meaningful conversations in our institutions.

An instructional designer's job is to assist the instructor in the assembly of a course. This includes the technical, as well as the strategic, construction of the course. The final product of the collaborative effort between the instructor and the instructional designer should be a well-tuned course that provides a carefully scaffolded sequence of course materials, assignments, and assessments that lead all students to a successful achievement of course objectives. In short, the course should be UDL ready, and this begins with the development of the syllabus.

Research has shown that students forget almost all course content two years after the end of a course (Bacon & Stewart, 2006). However, the reality of memory is not that simple. It might be better to say that students forget what they feel they don't need, what they don't need when doing significant activities, or what they don't care about. So, how do we put this all together and address all of these concerns? We need to do things differently than we have been doing.

To help us demonstrate this conversion process we will use a sample online course with a standard non-UDL syllabus and go step by step through the syllabus to make it UDL ready. The course is called "Teaching with Technology" and is for students who are interested in teaching online.

When you read the sample syllabus, you may note a number of similarities to your own course syllabus. Worry not. Throughout this chapter, we will provide step-by-step scaffolding so you can create or revise your syllabus to make it more universally designed. This will be the first step to designing and delivering a universally designed online course. In this discussion, we will focus on the following sections of a syllabus: instructor information, course description, course objectives, required reading, policies, assignments, and course schedule.

These components are also important to include when designing a virtual snow day, as well. Although the syllabus will not be as extensive, it's important to outline the expected outcomes of the online session and organize all required reading, assignments, and due dates in one place.

▶ Teaching with Technology

Poor Richard's University
Intersession
Tom Thibodeau

- Email: tthibodeau625@gmail.com
- Office Telephone: 102-555-0115
- Office Hours: T,R 7.15–8.00 pm and by appointment

Course Description

This course will provide an introduction to the theory and practice of teaching with common computer and network technology resources; its design is based on the understanding that many information technologies have multiple capabilities that can support a wide range of instructional pedagogies. As such, the emphasis of this course will be on how your current pedagogical strategies can be extended (and perhaps even enhanced) through the use of information technologies, rather than on adapting one's pedagogy to meet technology-based requirements.

Course Objectives

- Assess the pedagogical opportunities associated with current and emerging information technologies.

- Explore how information technologies can facilitate greater course interaction, cooperation, and collaboration.

- Experiment with information technologies to develop strategies that could help you better meet the diverse learning needs of your students.

- Leverage information technologies to enable you to achieve greater classroom and workflow efficiencies.

Required Readings

- There are no textbooks to purchase for this course (given the rapid pace of technology change, most traditional print-based texts are out of date prior to their distribution).

- All required readings will be available online.

Assignments
Discussion Board

- By Sunday at midnight: complete all readings and post your initial thoughts, questions, or comments on the discussion board. Must be a minimum of 200 words. Postings less than 200 words will not receive more than 3/6. Postings less than 100 words will receive a 0.

- By Tuesday at midnight: Post *at least two* thoughtful follow-up comments to the discussion board. Both comments must be at least one paragraph in length. Aim for a minimum of 100–150 words in each response. Responses less than 100 words will receive a 0.

Final Individual Portfolio
Your final project will contain representative work that you've produced this session, along with a reflective narrative on your pedagogy, new sample lessons that incorporate computer or network technology, and an annotated bibliography.

Final Group Project
The group project is relatively straightforward: all the members of the class contribute to a wiki that can be used by instructors who are interested in using technology in their teaching. The content of this wiki is open—it can include pedagogical considerations; applications of specific technologies; methods

of generating digital content; strategies for promoting online community formation; reviews of important articles/research—in short, anything that touches on the use of digital technology and teaching.

Policies

Attendance Policy

This classes is online-based; however, it is necessary for students to submit work EVERY week of class because attendance is based on whether or not work was completed that week. After two weeks with no submitted work, the student's final grade will be reduced by one letter grade (10 points). For each additional week without submitted work, the student will lose an additional 10 points. Ultimately, if the student is marked absent for 40% of the classes (four weeks without submitting work), the student will receive a failing grade for the course.

Academic Honesty Policy

Any assignment, project, paper, or examination is expected to be the student's own work, in the student's own words. Willful academic dishonesty will not be tolerated and may result in course failure. This includes:

- copying another student's work or allowing one's own work to be copied;

- using notes or books during an examination without the instructor's advance permission; or

- presenting information or images copied from a book, journal, or online source as one's own.

Course Schedule

Week 1: Asynchronous Communication Strategies. Primary question of the week: Why do we need to communicate at all with our students?

1. Email and listservs

2. Discussion boards, blogs (with and without subscriptions)

3. Wikis

Week 2: Content Management. Primary question of the week: How do we efficiently manage online content?

1. Online document publication (standards and best practices)

2. Document archives and document access

3. Managing multiple course sections/managing multiple platforms

Week 3: Synchronous Communication Strategies. Primary question of the week: Is this worth all the effort that it will take?

1. Virtual classrooms and whiteboards

2. Webinars, Google+, Skype, and other live multimedia communication tools

3. Optional webinar (synchronous video conference) will be scheduled this week

Week 4: Social Media Interaction. Primary question of the week: How do we leverage the tools that our students use on a daily basis?

1. Communicating with Twitter, Facebook, and so on

2. Using social media tools for course instruction and collaborative interaction

Week 5: Multimedia Content. Primary question of the week: How can multimedia help my pedagogy?

1. Creating simple audio and multimedia content using Audacity and/or Screencast-O-Matic

2. Posting and making multimedia content accessible

3. Multimedia instructional design (standards and best practices)

Week 6: Writing and Research Tools and Strategies. Primary question of the week: What writing and research tools can we use in our online teaching?

1. Feedback and response strategies

2. Mind mapping and idea-generation strategies

3. Reference and research tools

INSTRUCTOR INFORMATION

#step1: Make yourself come alive in the syllabus to be "present" on the page.

Let's get started with something easy, yet tremendously important in an online course—making it personal and human. In an online course, it is very easy for a student to think, "There is no one at the other end of this course," since much of the course is done independently through text and email communication. We are not trying to re-create an impersonal correspondence course.

You want to make sure that your students know who you are, what you look like, and what you expect from them in your course. So, you need to make sure that your students know "you" are in the course. Start by adding your picture and filling out the profile section of your learning management system (LMS) and ask your students to do the same. In most LMSes, the system will use your image in all communications and discussion board posts.

Many faculty who we have worked with are uncomfortable with using their own picture in the LMS, but we all need to realize that it is a very easy way to reinforce our role in and connection to the course. Many faculty (and students) like to use some sort of avatar or other image instead of a personal photo. Any image is better than no image, but the best image that you can use is your own. It does not need to be a close-up shot; it does not need to be a formal shot. It can even be a selfie! It should be a shot that best represents who you are and why you are teaching this course. It will help you establish your presence in the course.

Now, you have to "replace" yourself in this online course and still maintain a UDL-centered course. To do this, you have to create your instructor presence. Be very "present" in your course from the beginning when you design your syllabus. You want your personality to show so students can make a connection with you off the page of the syllabus. A short biography will help to strengthen your presence in the course for your students, and don't just provide your professional bio. Allow students to see you as a real person by adding some personal information. It's up to you what you're willing to share, but a little goes a long way.

We also recommend using a video link to introduce yourself in your course syllabus. While you are creating a video introduction, also link students to a video that walks them through the course syllabus. This video will be a "formal" orientation to the course for your students. Cover the major operational features of the course, and the operation of the LMS. This orientation could also include a review of the syllabus. These multiple representations are important for students, and as we say in UDL, "essential for some, good for all." This video needs to get to the students independent of the LMS. Plan on sending a link to the video introductions as part of a "week zero" welcome email before the start of the course. This will help students overcome the technical barriers of entering the course, as well as the emotional barriers, including the possible fear of taking the course.

After you provide some basic contact information and introduce yourself via video, make sure your students know how they can contact you and what your availability is for this course. Just because you will be teaching an online course that is available all the time doesn't mean that you will be. Students need to know how and when you will be available.

So, the first section of our syllabus should be something like this:

▶ **Teaching with Technology**

Poor Richard's University
Instructor: *Tom Thibodeau*

Click my name above to watch my short video introduction. For a video explanation of this syllabus, click *here*.

- Email: tthibodeau625@gmail.com

- Office Telephone: 102-555-0115

- Office Hours: T,R 7:15–8:00 pm and by appointment. Meetings can happen over the phone, via Skype, or in person if we happen to be geographically close.

- The best way to contact me is through email. I check my email very often and will respond within 24 hours, usually within less time. If I am unavailable for some reason, you will receive an out-of-office response that will tell you when I will return your email.

Instructor Biography: Tom Thibodeau

I am an assistant provost and the director of the Faculty Resource Center. My responsibilities include faculty development, the use of academic computing resources, classroom A/V resources, online resources (especially Canvas), and a variety of administrative tasks. I started at this institution 25 years ago as an adjunct faculty member and progressed to an associate professor in video production. Along the way, I helped create and chaired a new Multimedia and Internet Communications program and worked as the college's first educational technologist. I also have 20 years of experience in the video production industry as a videographer and online editor.

I live in Seekonk, Massachusetts, but have spent a lot of time in Worcester, since that is my wife Kathy's hometown. Kathy teaches the 5th grade at the local elementary school and we have three children: Katie, a curriculum administrator who has four young kids and is a specialist in Universal Design for Learning and the Common Core; Lindie, a marketing consultant and interior designer who has two little ones (grandkids are the best!); and Jeremy, who is an architect working for a construction firm in Boston. We all love animals and all have at least one cat or dog. My son Jeremy got us all involved in the Cheetah Conservation Fund when he was 8 and we were known for a while as the "Cheetah Family" and have been on the Discovery Channel's "Animal Rescuers'" program (that's a real long story). When I have some free time I love to create stained glass panels for my home.

COURSE DESCRIPTION

#step2: Before you write a course description, ID the mega-standard—what do students NEED to remember in the future? #whatsmostimportant

Next, we need to make sure that our course objectives and goals are clear, achievable, significant, demonstrable, and measurable and that their significance is embedded in the course discussion. Fink's (2013) framework of Significant Learning provides a great model for course design and serves as a solid foundation for creating a lesson-planning guide with UDL Guidelines. Fink's model holds two assumptions: when we teach, we engage in two activities. The first is that, as instructors, we design a course based on a number of decisions we make about how we want the course to be taught. The second activity is that as we implement, or deliver, our course, we engage in student-teacher interactions. Both the course and our interactions impact student learning outcomes. In order to teach well, Fink argues, one must be competent in course design and teacher-student interactions.

From the very beginning, by redesigning the instructor information, we are setting the stage for valuable and meaningful teacher-student interaction, so now you have to make decisions about the content of the course.

Often, a syllabus begins with the course description. Students have likely already reviewed this description when they registered for the course, so it must contain the essence of the course. Before you write your course description, ask yourself the following question: **What do you want your students to remember two years after the end of the course?** Now, remember, students forget almost all course content after a two-year period, so if they only remember one concept, one *mega-objective*, as we call it, what would it be? For example, in our sample course, "Teaching with Technology," the mega-objective might be:

Two years after the end of the course, the student will be able to easily experiment with new classroom technology and decide how well that technology will fit into the overall process of the course and will align to best practices.

This objective is aimed at getting the student to be open-minded and courageous about the use of technology, to recognize that it is a process not a destination, that the "teaching" is more important than the "technology." It is also aimed at getting the students to be comfortable with the technology itself, and to care about the value of technology in teaching when it is chosen appropriately.

To achieve this mega-objective, the course will ask students to research best practices, discuss potential options, and try new technology in connection to their teaching practice. But, since technology will always keep changing, it will be more important for students to learn how to evaluate the effectiveness of the technology than it will be to master the use of particular technologies. In essence, they need to learn how to care about what technology can bring to the educational environment. This mega-objective should be the essence of the course description.

To revise the course description so it's more UDL-centered, reflect on the following UDL Guidelines:

- *Highlight patterns, critical features, big ideas, and relationships*: In the course description, it's important to highlight the big ideas and keep the description simple. Why are you teaching the course?

- *Guide appropriate goal-setting*: Students need to know exactly what they will learn throughout the course. Return to your mega-objective and incorporate that into the course description so it's clear.

- *Optimize relevance, value, and authenticity*: In the course description, make the connection to why the course is valuable for students.

See how the description below incorporates the suggestions outlined in the UDL Guidelines identified above and focuses only on the mega-objective in the course.

BEFORE	AFTER
This course will provide an introduction to the theory and practice of teaching with common computer and network technology resources; its design is based on the understanding that many information technologies have multiple capabilities that can support a wide range of instructional pedagogies. As such, the emphasis of this course will be on how your current pedagogical strategies can be extended (and perhaps even enhanced) through the use of information technologies, rather than on adapting one's pedagogy to meet technology-based requirements.	This course is designed to provide opportunities for educators to experiment with instructional technology and assess which technologies are most valuable for student learning; the course design is based on the understanding that technology supports teaching and learning and is used to enhance best practices.

In the description above, we eliminated jargon and embedded our mega-objective. When examining your own syllabus, it may be helpful to share your course description with someone who knows nothing about your content. Ask him or her what the mega-objective, or overall purpose, of the course is. Do they hit the nail on the head or do they miss the point? This activity will help you to continually refine your course description so it is more universally designed.

COURSE OBJECTIVES

> **#step3:** Course objectives must show how students will express meaningful knowledge.

The next step is to revise the course objectives. Learning objectives need to be modified so they include options to meet learner variability, options for representing content, and options for student actions and assessments. This will require instructors to think globally about all students and create a course that will guide all students to the course outcomes. The course objectives are going to direct everything you want the students to do in your course. They have to be measurable and include how you will assess achievement of the goal. To do this, you need to define the options for the assessments or define them with flexible modes of expression in mind. For example, "prepare a response" is much more flexible than "write a 350-word essay," because students could respond using multiple formats. Having the expressive element embedded in the goal is crucial, because "it is not enough for students to acquire information; they must also have some way to express what they have learned, and some way to apply that information as knowledge. Only in its expression is knowledge made useful" (Rose et al., 2006).

So, our course objectives need to change from:

AT THE END OF THIS SESSION, YOU WILL BE ABLE TO:	AT THE END OF THIS SESSION, YOU WILL BE ABLE TO:
Assess the pedagogical opportunities associated with current and emerging information technologies;	Assess the pedagogical opportunities associated with current and emerging information technologies as evidenced by your active participation in weekly discussions, assignments, and a group final project;

AT THE END OF THIS SESSION, YOU WILL BE ABLE TO:	AT THE END OF THIS SESSION, YOU WILL BE ABLE TO:
Explore how information technologies can facilitate greater course interaction, cooperation, and collaboration;	Explore how information technologies can facilitate greater course interaction, cooperation, and collaboration by participating in, designing, and creating activities that improve interaction in your proposed courses;
Experiment with information technologies to develop strategies that could help you better meet the diverse learning needs of your students; and	Experiment with information technologies to develop strategies that could help you better meet the diverse learning needs of your students by using these technologies in this course and in your proposed courses; and
Leverage information technologies to enable you to achieve greater classroom and workflow efficiencies.	Leverage information technologies to enable you to achieve greater classroom and workflow efficiencies in the completion of your final project.

COURSE POLICIES

#step4: Highlight "big ideas" regarding course policies that may impact how students achieve in the course.

After you have written or revised your course objectives, you must make sure that your syllabus has all the information that your students need to be able to successfully navigate the course, overcome any technical difficulties, and understand all of the school and course policies that relate to the online education process. What is your policy on late work? What is the Academic Honesty Policy? Will you use a service such as Turnitin or SafeAssign to check the originality of your students' work?

All of these items must be detailed on your syllabus or you should link to the appropriate information on the web. Furthermore, you should require acceptance of these policies and procedures from your students by means of an acknowledgment form, a simple assignment, a web scavenger hunt, or a syllabus quiz because this will help to clarify the structure of the course and will allow students to acquire important background information that will eliminate possible barriers. If you are preparing for a virtual snow-day system, these policies must be in place and understood by the students before the snow day.

Your syllabus should include something that looks like this (adjusted to what is appropriate for your course, institution, or virtual snow-day needs):

Attendance Policy

This class is online-based. However, it is necessary for students to submit work EVERY week of class because attendance is based on whether work was completed that week. After two weeks with no submitted work, the student's final grade will be reduced by one letter grade (10 points). For each additional week without submitted work, the student will lose an additional 10 points. All assignments are expected to be completed in a timely fashion; however, we certainly understand that life contingencies occasionally arise, so just let us know as soon as possible if you need additional time to complete an activity.

Academic Honesty Policy

Any assignment, project, paper, or examination is expected to be the student's own work, in the student's own words. Willful academic dishonesty will not be tolerated and may result in course failure. This includes:

- copying another student's work or allowing one's own work to be copied;

- using notes or books during an examination without the instructor's advance permission; or

- presenting information or images copied from a book, journal, or online source as one's own.

All submitted work in this course will be reviewed through Turnitin for originality.

Academic Support

Academic support services are available through the Academic Skills Center, Student Support Services, and the library, and are available online or via email. The Writing Center and English Lab (located in the Academic Skills Center) are great places to get help with your writing assignments. The professional tutors help you with all phases of writing. While the professional tutors in the English Lab can help you in a variety of ways with all phases of writing, they cannot write the paper for you. They are not a proofreading or typing service. And they can't help you turn out a finished assignment 20 minutes before it is due. So, be considerate. Help them help you by bringing or sending a copy of the assignment sheet, the rubric, and your research materials with you.

Time Commitment

- This is a one-credit course. Assume approximately 3 hours per week of active work in the course.

- All but one activity will take place in an asynchronous fashion. In other words, you will be able to participate in most activities at a time convenient for you.

- There will be multiple scheduled opportunities to participate in the synchronous activity.

Participation Expectations

- Participation is demonstrated through regular and thoughtful interaction on the discussion board, and through blogs, email activities, and group projects.

- Completion of a variety of activities and assignments will be completed individually and in groups, as defined in the weekly course outline.

- Discussion board/blog posts should demonstrate some engagement in the discourse. For example:

  Really Bad: "Ditto."

  Not So Good: "Great thought, Jim!"

  Much Better: "Jim, I thought that your analysis of the 'no significant difference' debate missed an important point: that technology-based teaching strategies assume a certain level of economic resources. While I agree with you, in that technology-based pedagogies..."

Due Dates and Disaster Recovery

- Unless otherwise indicated, all assignments must be completed on the published due dates.

- Plan ahead and make local copies in Microsoft Word or another format (for example, Google Docs) of all materials submitted or posted online in the event of system failure.

- If you encounter specific hardware or network problems that prohibit you from completing an assignment on time, contact the instructor as soon as possible via email or telephone.

- If you experience recurrent technical problems that prohibit you from completing multiple assignments, you may be asked to re-enroll in the course at a future date.

Technology Expectations

- Regular access to a computer

- Broadband Internet connection

- Ability to send and receive email and email attachments

- Microsoft Office, Google Docs, or OpenOffice (see http://www.openoffice.org) applications. Adobe Reader (available free from http://www.adobe.com)

Technical Support

We can offer support for navigating our course website and getting the most out of our course and the tools that are used in your course. We do offer 24/7/365 phone, email, and chat support for our Canvas Learning Management System. Canvas help is available from the help menu at the top right of the Canvas application.

Acknowledgment

Please go to the Week 1 course module and complete one of the options to acknowledge that you have read this syllabus and understand the requirements of this course.

COURSE READINGS

#step5: Course content needs to be UDLified. #ithinkwecoinedthatterm

Every course needs reading assignments, and since you are creating an online course, you will need to provide links to much of the reading online. Arum and Roska (2011) presented very compelling research on the increased academic achievement of students who read more than 50 pages a week and write more than 20 pages a semester. But just because reading helps students learn and you have chosen the readings and posted the links doesn't mean that your students will actually read them. Research suggests that students see a weak relationship between course readings and academic success, and therefore, most college students do not read course assignments (Hobson, 2004).

You are going to have to do something very proactively to get the students to read—provide them with interesting and significant work that connects the course to the real world, provide alternatives to the content in video formats (from YouTube and other sources), and most importantly, have the students integrate what they learn from the readings (and the rest of the content) in everything else that they do in the course. Don't make them just read the content—make them use it.

To reduce any barriers to the process of reading, your course should also provide links to text-to-speech options for students who would like or need to access the information in audio format.

COURSE CONTENT

Now it is time to define the course content; what the students will be doing during the course. As you begin to develop this portion of the syllabus, try to think of what your students will be doing before you start to define the content they will be using and how you will help them do it. Try to identify potential barriers to the objectives in the online environment and ways that you will offer options to overcome those barriers. Will all students be able to achieve these goals regardless of your representations and their expression and engagement? It is imperative to consider possible barriers before designing the course, since "Universal design focuses on eliminating barriers through initial designs that consider the needs of diverse people, rather than overcoming barriers later through individual adaptation" (Rose et al., 2006). Building on Table 2-1, let's add a column for those barriers and a column for solutions (Table 3-1).

TABLE 3-1 Relationship between UDL Principles, Barriers, Brain Networks, and Solutions

TO ACTIVATE THE BRAIN: ADHERE TO UDL PRINCIPLES →	TO ADHERE TO UDL PRINCIPLES: IMPLEMENT GUIDELINES→	POTENTIAL BARRIER	TO IMPLEMENT THE GUIDELINES: IMPLEMENT THE FOLLOWING CHECKPOINTS	POTENTIAL SOLUTION
Provide multiple means of engagement.	Provide options for self-regulation.	Unclear expectations. Impersonal presentation of course objectives. Few opportunities for grades or feedback.	Promote expectations and beliefs that optimize motivation. Facilitate personal coping skills and strategies. Develop self-assessment and reflection.	Provide precise expectations that define success. Personalize all presentations and content delivery. Provide multiple opportunities for formative assessments and feedback—with or without grades.

TABLE 3-1 Relationship between UDL Principles, Barriers, Brain Networks, and Solutions *CONTINUED*

TO ACTIVATE THE BRAIN: ADHERE TO UDL PRINCIPLES →	TO ADHERE TO UDL PRINCIPLES: IMPLEMENT GUIDELINES→	POTENTIAL BARRIER	TO IMPLEMENT THE GUIDELINES: IMPLEMENT THE FOLLOWING CHECKPOINTS	POTENTIAL SOLUTION
Provide multiple means of engagement (continued).	Provide options for sustaining effort and persistence.	Overly general goals. Expectations of a one-size-fits-all goal. No group work or collaboration. Multiple-choice quizzes and exams only.	Heighten salience of goals and objectives. Vary demands and resources to optimize challenge. Foster collaboration and community. Increase mastery-oriented feedback.	Align objectives and assessments. Provide multiple options for each assessment. Expect and plan for collaborative group work. Provide more feedback than just grades.
	Provide options for recruiting interest.	Use a defined list of activities. Reuse the same activities for every group.	Optimize individual choice and autonomy. Optimize relevance, value, and authenticity. Minimize threats and distractions.	Engage students in the selection of the activities. Allow students to select their own groups some of the time.

TABLE 3-1 Relationship between UDL Principles, Barriers, Brain Networks, and Solutions *CONTINUED*

TO ACTIVATE THE BRAIN: ADHERE TO UDL PRINCIPLES →	TO ADHERE TO UDL PRINCIPLES: IMPLEMENT GUIDELINES→	POTENTIAL BARRIER	TO IMPLEMENT THE GUIDELINES: IMPLEMENT THE FOLLOWING CHECKPOINTS	POTENTIAL SOLUTION
Provide multiple means of representation.	Provide options for comprehension.	Deliver content in only one mode. Assume that all students understand all resources. Never explain the reasons for the choice of your course content. Handle all resources in silos.	Activate or supply background knowledge. Highlight patterns, critical features, big ideas, and relationships. Guide information processing, visualization, and manipulation. Maximize transfer and generalization.	Provide text, audio, and video of all resources. Explain each resource with an introduction, and connect each resource to the next with clear language.
	Provide options for language, mathematical expressions, and symbols.	Assume that all students understand the words and symbols you use. Provide only one explanation of a resource, if any. Ignore language differences within your course. Limit use of images and graphics.	Clarify vocabulary and symbols. Clarify syntax and structure. Support decoding of text, mathematical notation, and symbols. Promote understanding across languages. Illustrate through multiple media.	Provide a glossary for each module or section. Provide links to additional and varied supporting information. Survey your class for primary languages, and support if possible. Use lots of images, graphics, and video.

TABLE 3-1 Relationship between UDL Principles, Barriers, Brain Networks, and Solutions *CONTINUED*

TO ACTIVATE THE BRAIN: ADHERE TO UDL PRINCIPLES →	TO ADHERE TO UDL PRINCIPLES: IMPLEMENT GUIDELINES→	POTENTIAL BARRIER	TO IMPLEMENT THE GUIDELINES: IMPLEMENT THE FOLLOWING CHECKPOINTS	POTENTIAL SOLUTION
Provide multiple means of action and expression.	Provide options for executive functions.	Set a due date and never remind students. Never require students to "manage their own group activities."	Guide appropriate goal-setting. Support planning and strategy development. Enhance capacity for monitoring progress.	Scaffold all assignments. Create incremental due dates. Require multiple check-ins with project progress. Require multiple collaborative activities.
	Provide options for expression and communication.	Allow only one mode of fulfillment.	Use multiple media for communication. Use multiple tools for construction and composition. Build fluencies with graduated levels for support for practice and performance.	Allow for maximum flexibility of student response in text, audio, video, or kinesthetic activity whenever possible. (Kinesthetic activities will always be recorded video online.)

TABLE 3-1 Relationship between UDL Principles, Barriers, Brain Networks, and Solutions *CONTINUED*

TO ACTIVATE THE BRAIN: ADHERE TO UDL PRINCIPLES →	TO ADHERE TO UDL PRINCIPLES: IMPLEMENT GUIDELINES→	POTENTIAL BARRIER	TO IMPLEMENT THE GUIDELINES: IMPLEMENT THE FOLLOWING CHECKPOINTS	POTENTIAL SOLUTION
Provide multiple means of action and expression *(continued).*	Provide options for physical action.	Lecture without interruption.	Vary the methods for response and navigation. Optimize access to tools and assistive technologies.	Use the "15 minute rule"— require a different action every 15 minutes. Provide multiple means of support in and out of your course.

Once you have identified the barriers to student success, you can provide flexibility of delivery. This will lead you to the use of multiple modes of representation in the presentation of your course content, assessments, and feedback. It will allow you to be creative and open to innovation and give you the ability to promote multiple modes of expression. And, since you are already working online you will have a variety of digital solutions available for you to use. The world of Web 2.0 tools awaits.

Web 2.0 tools are digital tools that are built in the "cloud" and are social in nature, usually easy to use, and are available at low or no cost to you and your students. They are built for collaboration. They offer a wide variety of ways to manage text, audio, images, video, and the ideas the create them and promote them. Tools like Google Docs, WordPress, and Evernote are no longer just word processors, but collaborative engines that support multiple students who work together as they develop their ideas and finalize their projects. YouTube, iTunes, and Vimeo become more than just video streaming services; they become integrated components of your course that allow you to deliver content resources to your students. Skype, Google Hangouts, and the BigBlueButton allow for more than just a real-time, face-to-face discussion; they provide a real-world opportunity to live and work in a truly global workplace.

DISCUSSION BOARDS AND BLOGS

Discussion boards and blogs usually make up a considerable portion of an online course or virtual snow day. Discussion boards are important, but should not be considered the only way for you to create interaction between your students and the content, your students and each other, and your students and you. But, let's start here. As you remember, in the previous chapter we looked at revising the description of the online discussion expectations so they are aligned more closely to the UDL Guidelines.

ASSIGNMENTS AND ASSESSMENTS

You will also need to change the way you design assessments. The simple fact that you will be allowing a variety of modes of expression from your students will require you to change your assessment philosophy. The easiest way to approach this is to change your style of assessment and do a mix of formative and summative assessments. Formative assessments are usually ungraded and provide you with information on the progress that each student is making in the course. Your LMS probably allows you to create quizzes, exams, and surveys. Each of these can be used for a formative assessment.

As we have stated before, online courses should not be a digital version of a correspondence course. Just because the course is online, it doesn't mean that you can't ask students to work together. You need to build true interaction between you and your students, and your students and each other, in order to foster collaboration and community within the course. A great way to do this is with projects and group work. You have a lot of options here. You don't have to do group work each week, but you do have to scaffold the process so that students will know how to successfully complete the assignment and work together at the level you expect.

You can also make your final project a group project. Being able to work in a team is a very sought-after skill in the global marketplace.

If you are presenting your assignment guidelines in the syllabus, or within specific course modules or instructions for a virtual snow day, provide detailed scaffolding for each assignment and be sure to encourage multiple tools for construction and composition. Providing specific details will support planning development, and guide appropriate goal setting for students. If you examine the description of the final portfolio assignment below, it's not clear what students are expected to hand in, so it must be revised so it's clear what is expected, but also that flexibility is encouraged.

Before	Your final project will contain representative work that you've produced this session, along with a reflective narrative on your pedagogy, new sample lessons that incorporate computer or network technology, and an annotated bibliography.
After	*Introduction*
	Your final project will contain representative work that you've produced this session, along with a reflective narrative on your pedagogy (either in written, video, or audio form), new sample lessons that incorporate computer or network technology, and an annotated bibliography. Sample projects and assignment rubrics are posted in the LMS under "Assignments."
	Please include the following items in a single point of access (for instance, in a single Microsoft Word file or Google Doc, a website, etc.).
	Contents
	Develop a reflective narrative on the relationship and role of computer and network technologies on your teaching philosophy and practice. This narrative should provide a relatively broad perspective and can be presented from the first person point of view (if desired), using relatively informal language. This narrative can be written or recorded using audio or video.
	Produce an annotated bibliography that contains 10 items that would be of interest to others who want to incorporate information and network technologies in teaching their specific disciplines. Please use APA format. A sample is posted in the LMS.

After (continued)	*Draft three sample lessons* that leverage information, network technologies, or both, along with a brief discussion of the specific objectives associated with the technology use. You may format the lessons in any way you choose. Sample lesson formats are available in the course LMS under "Final Project/lesson plans."
	Produce sample video, audio, or web content that integrates into one of your lesson samples and uses the most appropriate technology that services your lesson. In a simple reflection, include a link to your content and discuss the design of your content (approximately 250–500 words, but it can be longer if you wish). Again, this discussion can be written or recorded using audio or video.

COURSE SCHEDULE

The course schedule will outline the "meat" of your course. It will provide the resources, the sequence, and the assignments, as well as all of the dates and deadlines. Choose your content wisely and remember that access does not equal learning. As Rose et al. (2006) remind us, "Making information accessible is not enough. The goal of education is not only to make information more accessible; that is a goal for librarians, publishers, or engineers of popular search engines. The goal of education is to teach students how to work with information, including finding, creating, using, and organizing information. There is an important distinction between accessing information and using it" (p. 3).

It is very important that you create multiple modes of representations for the same content in your weekly modules. This means that ideally you would have text, audio, video, and computer variations for your content. Since the course is online just about everything you do will be computer-based, so it will be important to find ways to bring the other modes into the course. You do not have to create all options for all items, but you do want to provide enough variation to give everyone a choice every week. You also can build these resources as you continue to build your course or lesson each time you prepare to teach it.

Each week you should also plan for individual work and group interaction in discussion boards, blogs, wikis, and other assignments. This will require detailed instructions on each assignment. We like to build these assignments sequentially over several weeks in the course. You will need to plan appropriate scaffolding of concepts and activities from lesson to lesson. "For example, scaffolds and supports at the postsecondary level can include review sessions, opportunities for students to receive feedback on project topics before they are submitted, and optional readings to address learners with different levels of prior knowledge (i.e., readings providing either background information or advanced discussion of course topics)" (Rose et al., 2006, p. 4).

You don't necessarily have to construct all of this yourself. You can find proper support and resources for each activity on the web. Just plan to communicate specific instructions for the different options each week. In a recent course, we had a situation where we asked students to write a personal statement, not a research paper, for a weekly assignment, but 60% of the students submitted a research paper. It turns out that there was a college policy (listed in an early pre-course introduction) that encouraged all assignments to be submitted as a research paper, and our instructions for the personal statement were not specific enough.

Plan for a variety of synchronous and asynchronous activities throughout the course. The flexibility and convenience of taking an online course does make synchronous activities a bit harder to schedule, but it is usually worth the effort. You may have to offer two or three synchronous sessions to fit everyone's schedule, but if you can only do this once per session, you will see a positive effect on the sense of community in the course. Many LMSes have built-in tools like BigBlueButton (http://bigbluebutton.org/) that let you offer group webinars or video conferences or even online office hours.

There are other options on the web as well: Free Conference Call offers free services for both group phone calls and online meetings (https://www.freeconferencecall.com). But if this is unavailable to you, you can use Skype for one-on-one video calls, an LMS chat feature, or even a shared Google Doc that everyone can edit.

If you want your students to view a video, you can use a video annotation service like http://corp.hapyak.com/ to focus and assist your learners in the process of understanding the video. HapYak allows you to add annotations, divide the video into chapters, add links to supporting information, and create video assessments. HapYak allows for the creation of five interactive videos per month for free. Also note that you can use the transcription services of YouTube for videos under 10 minutes to provide a text version of your videos. Please review the accuracy of the of transcript for errors before you release it to the students.

Once you have considered how you will organize your course, revise your weekly schedule so students can see how you've aligned your course to UDL. Here is an example of what each week would look like.

▶ Course Schedule/Outline

Week One (October 1, 2016): Asynchronous Communication Strategies: Please see the week one module for more details.

1. Primary question of the week: Why do we need to communicate at all with our students?

    - Email and listservs

    - Discussion boards, blogs (with and without subscriptions)

    - Wikis

2. Reading:

    - "Disrupting Ourselves: The Problem of Learning in Higher Education," by Randy Bass: http://www .educause.edu/EDUCAUSE+Review/EDUCAUSEReview MagazineVolume47/DisruptingOurselvesTheProblemo/ 247690

    - "How Disruptive Innovation Changes Education Q&A," by Martha Lagace with Clayton M. Christensen, Michael

---

B. Horn, and Curtis W. Johnson, August 18, 2008: http://hbswk.hbs.edu/item/5978.html

- "So You've Got Technology. So What?" by Richard A. DeMillo: http://chronicle.com/article/So-Youve-Got-Technology-So/131663/

- "Disruptive Education Technology: Helping Kids Learn" by Michael Horn: http://images.businessweek.com/ss/08/10/1021_education_tech/index.htm

- "Implementing the Seven Principles: Technology as Lever" by Arthur W. Chickering and Stephen C. Ehrmann: http://sphweb.bumc.bu.edu/otlt/teachingLibrary/Technology/seven_principles.pdf

3. Videos: These videos are an alternative to the readings above.

- "Learning, Innovation and possible futures of Georgetown" by Randy Bass: http://vimeo.com/68049901

- "Dr. Clayton Christensen discusses disruption in higher education" by Clayton M Christensen: https://www.youtube.com/watch?v=yUGn5ZdrDoU. Watch around the 50:00-minute mark for an interesting discussion about online learning and teaching individual students.

4. Activities

- Post to the discussion board—Introductions: Post a personal introduction in this forum as soon as convenient. Please see the week one module for more details on how to navigate the discussion board.

- Post to the discussion board—Instructional Technologies—A Starting Point: Initial post by Saturday, noon; three follow-up posts due no later than Tuesday, noon. Please see the week one module for more details on how to navigate the discussion board.

- Email Posting on Wikipedia to the listserv (send assignment to COWC-TWT@yahoogroups.com): no later than Tuesday, midnight, though sooner is better. If you need help with the listserv check out this link: https://info.yahoo.com/privacy/us/yahoo/groups/details.html

5. Need help?

- Tutors at the Academic Skills Center (ASC) are available to help you improve your reading ability or your writing skill. Call 202-555-0115.

- Need a text-to-speech app? Please check out these links:

- https://digitaltext.wordpress.com/at-tools/text-to-speech-options/free-text-to-speech-options/

- https://www.apple.com/accessibility/osx/

After you get your "system" down, consider building a different-looking course syllabus. Embrace the UDL Guideline "customize the display of information" to give students a different representation of your original syllabus. For example, you could create a graphic organizer that outlines the course objectives and connects them to the guidance for the corresponding assignments. Use Prezi or PowerPoint to create a slide for each section of the syllabus. You could also create a website for the syllabus with all videos, online readings, and Web 2.0 tools embedded into the site.

Finally, review links, spelling, grammar, and all references in your syllabus to ensure that you're sharing a quality product with students. To review, use Table 3-2 to make sure your syllabus is aligned to the Guidelines.

TABLE 3-2 Revised Syllabus Aligned to the UDL Guidelines

SYLLABUS SECTION	CORRESPONDING UDL GUIDELINES
Instructor Information	• Foster collaboration and community • Offer alternatives for auditory information • Offer alternatives for visual information • Illustrate through multiple media
Course Description and Objectives	• Heighten salience of goals and objectives • Highlight critical features and big ideas • Guide appropriate goal setting
Policies	• Facilitate managing information and resources • Minimize threats and distractions • Facilitate personal coping skills and strategies
Assignment Guidelines	• Heighten salience of goals and objectives • Vary the methods for response and navigation • Use multiple tools for construction and composition • Build fluencies with graduated levels of support • Support planning and strategy development • Facilitate managing information and resources • Enhance capacity for monitoring progress • Optimize individual choice and autonomy
Course Schedule/ Outline	• Offer alternatives for auditory information • Offer alternatives for visual information • Illustrate through multiple media • Use multiple tools for construction and composition • Facilitate managing information and resources • Optimize individual choice and autonomy • Optimize relevance, value, and authenticity • Vary demands and resources to optimize challenge

To conclude this chapter, let's take a look at our five students and how these syllabus revisions would help them to access the course from the start.

WHAT THE STUDENTS NEED FOR SUCCESS	HOW THE SYLLABUS MEETS THOSE NEEDS
Coco does not have a lot of time to spend online on virtual snow days because her two siblings have to work online, as well.	Coco would benefit from a clear syllabus or instructions for the virtual snow day that outlines all her options. If she knows that she can record her voice on her phone instead of typing a response, she will be much less anxious about completing all the work. Also, if she knows exactly how her work will be graded, she can assess her own assignments against the rubrics, which will make her feel better about handing the assignments in.
An online course for Kriti must find the balance between gathering information online, in books and articles, and in doing projects out in the "real world." Also, she needs support because she is nervous about taking classes in a virtual high school since she was not successful in a traditional setting.	Since Kriti has a difficult time paying attention, she will definitely benefit from the multiple representations of a syllabus. She can transition from reading, to watching videos, to viewing a PowerPoint. Also, it's clear how she can access support each week in the course schedule and also how she can contact the instructor or the Academic Support Center if she is falling behind.
To reach June, and address her challenges, we need to provide her with a warm and inviting online course that is easy to look at, easy to follow, and has lots of options. She will need lots of tutorials and support. Introductions need to present the activities of the week and be available as videos and text for her to print out and read again at her leisure. Assignments and activities need to be clearly explained in text and supported with examples of "good" work.	Our instructor introduction and video sets the stage for a warm and inviting environment. Also, it's clear from the course schedule that the course will have a lot of options for June to choose from. Since we offer everything in text, she can choose to print out the syllabus and all readings to read on her own time. Lastly, since we scaffold what is expected from assignments, and post exemplars, she knows what success looks like.

WHAT THE STUDENTS NEED FOR SUCCESS	HOW THE SYLLABUS MEETS THOSE NEEDS
José wants real projects that he can use to build his understanding of the instructional design process and build a portfolio that he can use to apply for a new job in the near future. José will need to be able to choose between multiple options for his assignments so that he can find just the right one for him to put his heart and soul into each week.	To engage José, we need to make a connection with him. Providing the instructor information will start to build this relationship. Also, in the assignment guidelines and course schedule, it's clear that the demands and resources are varied to help meet the needs of all students. José can push himself to excel within these expectations. Also, since the portfolio will include all assignments that have real-world application and can be used in a future online environment where he is the instructor, it will make his work more meaningful.
To help Ray, an online course will have to define all of the steps necessary to do the assignments and be interesting enough to challenge him and motivate him to find the time between work and family to do all the coursework. He will particularly need multiple forms of representation to engage him in the content. Since he is not a reader or a writer, the course will have to provide options for audio, visual and kines-thetic modes of representation and expression.	In the original syllabus, it's not clear how much time should be spent each week to be successful in the course or what types of activities will be expected. Basically, Ray would be going into the course blind. The improved syllabus outlines the time requirements, and the course schedule outlines the types of activities and how long he has to complete them. Knowing what is expected and knowing exactly how much time he will need to set aside is a good start. Also, in the attendance policy, the instructor acknowledges that life sometimes gets in the way, which will help to minimize the threat of failure for Ray.
	Unlike June, Ray is not a reader, so the video links and video and audio options in the course assignments and course schedule will motivate him.

The syllabus is just the first step, but it's an important one when designing a course. In the chapters that follow, we will continue discussion of course design, using the syllabus as the starting point, and then we will get into the magic—delivering the online course. When designed and delivered using UDL Guidelines and principles, we eliminate the barriers inherent in the less successful versions of online courses and we can increase student motivation and learning. With these changes, it is our hope that two years after the end of the course, our students will remember everything outlined in our course objectives.

REFLECTION

- Whether you teach an online course or only teach online for virtual snow days or flipped classes, it is valuable to design a syllabus to help students manage information and resources and monitor their progress. Take a moment to examine a syllabus for a course you teach. Review each section of your syllabus and judge the value of each section in terms of its accessibility to all students. What do you imagine the short- and long-term outcomes would be if you eliminated the barriers you identified?

OPTIONS FOR EXPRESSION

- Take the time to revise one or more sections of your syllabus, aligning the content to the research and recommendations presented in this chapter. You may complete this work independently, at a department meeting, or with your students. Consider posting the syllabus and encouraging students to comment on any sections of the syllabus that are confusing, unfair, anxiety-producing, and so on. Ask them to support their opinions with specific excerpts from the document. This will allow them to reflect on themselves as learners, optimize their motivation, and share mastery-oriented feedback with you.

- Customize the display of your syllabus. Record a video summary of your syllabus or create a graphic organizer, a Prezi, or a Power-Point for each section of your syllabus to share with students. You could also create a website or a Google Doc for your syllabus with all course materials and tools embedded into the site.

Cultivating "Instructor Presence" to Support Engagement

4

THIS CHAPTER OUTLINES the research on instructor presence and makes concrete connections to the affective networks (the areas of the brain responsible for motivation) and the UDL Guidelines that relate to engagement. A list of strategies is provided to support instructors in online environments when working with students.

Benjamin Franklin never ceases to amaze. We have already shared a pretty impressive résumé—building the foundation for the University of Pennsylvania, being instrumental in the development of the United States Postal Service, and inventing shock therapy. Just read your friends' Twitter profiles. They pale in comparison. And as if educator, community developer, and inventor wasn't enough, Poor Richard is also credited with making swimming a mainstream, "socially acceptable" exercise activity (Cleary, 2011). Yes, that's right. He pretty much invented swimming, too.

In *The Ethos Aquatic: Benjamin Franklin and the Art of Swimming*, Cleary shares that although swimming had been practiced for generations, it was thought to be a hazardous activity and was in fact prohibited in some areas. Franklin was so enamored by water and its pluripotentiality (you'll have to Google this one), he mentioned swimming explicitly in his *Proposals Relating to the Education of Youth in Pensilvania*, observing that all learners should be "frequently exercise'd in Running, Leaping, Wrestling, and Swimming."

Franklin was captivated by the strength and grace of water and felt that everyone should become a swimmer. When sharing advice on how to learn to swim, he noted, "[Y]ou will be no swimmer till you can place some confidence in the power of the water to support you; I would therefore advise the acquiring that confidence in the first place; especially as I have known several who by a little practice necessary for that purpose, have insensibly acquired the stroke, taught, as it were by nature" (Cleary, 2011, p. 60).

Franklin knew that water has the potential to support the swimmer— its buoyancy being necessary for success. We can extend this premise to an analogy of learning. In an online learning environment, for example, the instructor needs to be present, and design and deliver a course that allows all learners to be "buoyant" so they do not sink. The learners need to have confidence in the "power" of the instructor, but this is difficult to do when the instructor is not present in the traditional sense. This is why all instructors need to develop a presence early, so students can "acquire" knowledge "as it were by nature."

Once a properly scaffolded syllabus or detailed lesson plan is created, instructors also have to manage the course to maintain student engagement, monitor student progress, and provide mastery-oriented feedback to increase learning outcomes. This management process is most evident in the concept of instructor presence. Instructor presence is about positioning oneself in a course to increase student learning outcomes (Dennen, 2007). By simple definition, *instructor presence* is the quality of being "visible," or present, in online coursework. This includes sharing both your professional knowledge as well as your "personhood," including your personality, thoughts, and beliefs (Reupert et al., 2009). Although a simple definition exists, the reality is that instructor presence is much more complex, and an understanding of all the complexities will help you to be more visible in all aspects of your course. Instructor presence is a multifaceted construct.

The first week of a new online course is a critical time for establishing the instructor's presence (Dennen, 2007). Because there is no instructor at the front of a classroom, students will "get to know" the instructor first through the course syllabus or virtual snow-day plan, and then through any text, images, or videos presented early in the instruction.

The concept of instructor presence is closely related to research on instructor immediacy (Baker, 2010). *Immediacy* can be described as the psychological distance that a communicator puts between himself or herself and the object of his or her communication. In the case of an online course, the immediacy would be the psychological distance between the instructor and the students. A question to ask yourself when considering student perceptions of your immediacy would be, "Do students feel like they could get in touch with me quickly and easily and that I would be responsive?" To make sure the answer is yes, you need to minimize threats and distractions, be open to student questions and comments, and be responsive to all students, regardless of how they choose to contact you.

How can we, as instructors, build our presence? First, we need an understanding of the two internal constructs that make up instructor presence: social presence and cognitive presence (Hodges & Cowan, 2012; Ke, 2010).

> **@PoorRichardUDL** · What you would seem to be, be really.
> #socialpresence #aphorismsareawesome

SOCIAL PRESENCE

Social presence is how instructors communicate and interact with the students to provide mastery-oriented feedback on the discussion board and through emails, and by reaching out to students via the phone, video chat, or social platforms such as Twitter, Facebook, and so on. Research suggests that video provides more social presence cues than audio because of its ability to communicate facial expressions and intonation as well as other social cues. Audio provides more cues than text because it, too, includes vocal intonation. Common social cues are the use of humor, communicating emotions, self-disclosure, support or agreement for an idea, addressing people by name, complimenting another's idea, and allusions of physical presence (Rourke, Anderson, Garrison, & Archer, 2001).

These cues, used in video and audio, create a similar intimacy that is apparent in face-to-face interactions (Wise, Chang, Duffy, & Del Valle, 2004). To increase social presence, therefore, instructors can post video, audio, and text to differentiate the modes of representation, and also encourage students to do the same to increase the interaction between students, differentiate the modes of expression, and set the social presence tone for the online learning experience (Wise, Chang, Duffy, & Del Valle, 2004).

Injecting video, or facilitating video chats, can increase social presence and make the course similar to the traditional face-to-face classroom. As we noted previously, many instructors have not been trained in teaching an online course, and even if instructors have been trained, some simply don't enjoy the experience. Sharpe (2005), a community college instructor, discusses why online teaching turned her off: "I didn't know how to teach [students] to think online. I wanted to watch their faces while I talked; I wanted to hear their answers in real time; I wanted to challenge their replies, but with softness in my voice; I wanted my teacher fix: seeing a face light up with understanding." Seeing this light is possible with synchronous video learning experiences, but these experiences are unlikely if instructors aren't committed to modeling and building this aspect of their online course.

When instructors don't believe they can affect student learning in a virtual environment, it decreases their feelings of efficacy, which in turn can impact student achievement in their online course. When instructors have a high sense of efficacy, they set more challenging goals for students and persist despite obstacles to student learning (Ross, 1995). In general, teachers who believe that they can positively influence student outcomes are more likely to provide struggling students with support and the opportunities they need to succeed. This in turn will increase student motivation and engagement (Barkley, 2006). So, not only is it important to have a social presence, but instructors must also believe in the value of online education and their ability to deliver the content in a way that will allow all learners to succeed.

When instructors model social presence and have high feelings of efficacy, students contribute more and the quality of their interactions increases (Wise, Chang, Duffy, & Del Valle, 2004). In a study of social presence in an online graduate program for nursing, researchers measured social presence using the Social Presence Scale (Mayne & Wu, 2011). The items on the Social Presence Scale (Table 4-1) all relate to Universal Design for Learning (UDL), since agreement with the statements would suggest that the instructor built a learning environment that minimized threats and distractions, increased collaboration and community, and increased engagement. The results of the study suggest that the application of social presence cues by the instructor in online courses has a significant and positive effect on student perceptions of social presence and results in an increased desire to pursue future online education (Mayne & Wu, 2011).

> **@PoorRichardUDL** • Social presence paves road for persistence in education #doittoday

TABLE 4-1 Social Presence Scale Items (Mayne & Wu, 2011)

Online education is an excellent medium for social interaction.	I felt comfortable conversing in this course.
I felt comfortable introducing myself.	I formed a sense of participants' identities.
I was comfortable interacting with others.	I felt part of an online community.
I felt engaged most of the time.	I was able to form distinct impressions of participants.
I was able to form a distinct impression of instructor.	I felt a connected presence with group.
I felt a connected presence with instructor.	I felt like I participated in a shared learning experience.

COGNITIVE PRESENCE

Cognitive presence is the process of supporting students in collaborative inquiry so they can build knowledge (Garrison, Anderson, & Archer, 2000). To be cognitively present, instructors must shift from lecturing to questioning and probing. This helps students develop self-reflection and assessment skills. We recommend using the Paideia Seminar method in discussion boards and when providing feedback on assignments and assessments. The Paideia Seminar values the power of questioning in building shared knowledge, embodies the UDL Guidelines, and encompasses many important 21st-century skills. If you are preparing for a virtual snow day you will probably be preparing for one component for the Paideia Seminar unless you are preparing for a multiple-snow-day event.

What is the basic structure of a Paideia Seminar? A Paideia Seminar (adapted from Novak, 2014) includes a pre-seminar, the seminar, and a post-seminar and can be used in preparation for a discussion board posting, during the posting, and to reflect after the posting in order to build cognitive presence, one component of instructor presence.

Pre-seminar In this session, you will review standards and objectives, activate or supply important background knowledge for the text, clarify important vocabulary, and ask students to participate in a self-assessment. This can be done in electronic presentation formats including PowerPoint or Prezi, video, or audio (or all of them!). During the self-assessment, students reflect on how they usually participate in online discussions and create an individual goal for the session. You can ask students to post their goals. An example of a goal may be, "I will use at least three specific citations from the online readings for this week in my discussion." Also, be sure to review any agreed-upon rules for discussion each week to minimize threats and distractions, which would prevent students from participating.

Seminar ("Discussion") During the seminar, or online discussion, facilitate the discussion by asking open-ended questions when responding to students' posts. As students answer, require them to refer to text, video, or their classmates' ideas by citing specific details. Ask questions to encourage students to transfer their knowledge or discuss its applicability to their own learning.

Post-seminar In the post-seminar, ask students to reflect on their progress toward their individual goals and the goals or objectives of the week's discussion. This reflection can be done in multiple formats. For example, students could reflect by writing poetry, on the discussion board, or in a personal blog, sketching an image, recording audio or video, creating a podcast, and so on. Provide choices for students so they can reflect in ways that are meaningful to them.

Having students set goals for their learning based on a summary of key concepts, and returning to those goals and concepts after questioning, increases instructor cognitive presence. Research also suggests that summarizing key concepts before, during, and after instruction is a UDL strategy that has the most impact on changing instructors' behavior immediately following the training session when they are trained in that technique (Schelly, Davies, & Spooner, 2011). Also, student outcomes are increased when instructors assert and display their desire to engage in learner-centered discussion (Dennen, 2007).

Both social and cognitive instructor presence can be further discussed in terms of three components: instructional design, facilitating discourse, and direct instruction (Garrison, Anderson, & Archer, 2000).

All three components of the construct directly relate to UDL. The UDL Guidelines, when used proactively and deliberately, help instructors design a course where they are present in a virtual environment.

@PoorRichardUDL • UDL is the philosopher's stone. Be deliberate and turn learning to gold. #udlisgoodasgold

INSTRUCTIONAL DESIGN

Effective instructional design involves more than just an understanding of course design. It must involve both leadership and innovation. Innovation is necessary because education is constantly changing. To keep up with the times, an instructional designer must see and interpret trends in education and understand the changing needs of learners and the barriers they face: "By looking at the efficacy of a design with its potential shortcomings or barriers to learning, demonstrated by solutions that may or may not align with learning goals, an instructional designer assesses future learning potential" (Ashbaugh, 2013, p. 75).

In the hypothetical course "Teaching with Technology" presented earlier, the mega-objective speaks to the need for instructional designers to embrace technology and its ability to increase learning outcomes. Learning specific technologies is not important, since technology is constantly changing and the technological vehicles of the day may be extinct tomorrow (take MySpace for example). As research argues, "An instructional designer is challenged by a continually moving target in a fast-paced, changing world" (Ashbaugh, 2013, p. 79). Changes in education happen seemingly overnight, so all designers must develop courses for the future needs of students, and envision better solutions for a variety of learners; otherwise courses are at risk of becoming endangered.

Instructional designers, and instructors who are also instructional designers, must also be leaders. To be a leader, one must have the mindset of a leader, a vision, strong communication skills, the ability to collaborate with others, a strategy, and strong character (Ashbaugh, 2013). Mindset is important because instructors must believe that they have the skills to carry education into the future. One cannot learn about online education today and continue to teach the course in the same way five years from now. The research on best practices may not change, but the technology surely will. Having this mindset and feelings of efficacy are also important to developing a vision for the student experience in the online course. This vision, we must note, may not align to the vision of the learning institution, but being a leader empowers instructors to "resist an inflexible mandate and to give voice to a vulnerable population" (Ashbaugh, 2013, p. 80). This may mean giving a voice to those students who face barriers so that they can succeed in any online learning environment that is constantly changing.

Social presence and cognitive presence can both be discussed in terms of instructional design. Instructors must be innovative when developing their presence. They must remain curious and experiment with new technologies as they develop to strive to replicate the social and cognitive interactions that increase student satisfaction and learning outcomes in face-to-face classrooms. They must also be leaders in their practice to ensure that the needs of all types of students are met now and in the future.

FACILITATING DISCOURSE

Cognitive presence is built primarily through discourse with students, although instructors can also build social presence while participating in online learning discourse as well. In order to increase student learning, instructors must eliminate barriers to online discourse, yet many exist. Research suggests that the barriers to online discourse include confusing discourse coordination, incoherence of contributions, and problems with maintaining relevance in discussion threads (Pfister & Oehl, 2009). All of these barriers can be eliminated by implementing UDL Guidelines.

One barrier to online discourse is lack of discourse coordination, which can be addressed by encouraging students to use explicit referencing. When posting to a discussion or chat, instructors can model the use of making an explicit reference to a previous contribution or to instructional material (Pfister & Oehl, 2009). This modeling provides necessary scaffolding to students, supports their strategy development, and facilitates the management of resources, all of which are UDL Guidelines. This, in turn, helps to facilitate discourse and increase collaboration among students.

Research also suggests that instructors can better facilitate discourse by classifying and labeling all contributions with respect to their communicative type, for example, as a question, explanation, or critique (Pfister & Oehl, 2009). When trying to build cognitive presence, it's important that students know that your contribution is a question, so they can plan to respond and extend their learning. Students can also question each other and build cognitive presence among the community of learners.

For example, let's take June, our student in "Teaching with Technology." Imagine you want to contribute by asking a question, and you expect a response. You could write something like, "So June, you mentioned that you found the concept of social presence confusing. When you are on a discussion board do you feel any connection to me as the instructor or to your fellow students? I look forward to your response." After reading this comment, it would be clear to June that you expect her to continue the discussion by reflecting on her interactions in the course.

Facilitating discourse is important in online coursework because it targets student engagement. Instructors can facilitate discourse by drawing students in, encouraging their participation, encouraging the explicit reference and labeling of posts to organize the communication, and acknowledging and reinforcing student contributions throughout their discussions. It's important to note, however, that the instructor has to be present in the discussion without overly directing the conversation. It is the task of instructors to direct students to monitor their own progress and direct their own learning by being a guide.

@PoorRichardUDL · Discourse can help the student too. #letstalkitout

DIRECT INSTRUCTION

Direct instruction is a teacher-centered approach where instructors teach specific skills and concepts by providing detailed instruction, closely monitoring student progress, and offering frequent feedback (Lowe & Belcher, 2012). Direct instruction in an online course can take many forms: online lectures, summarizing discussions to confirm understanding, providing mastery-oriented feedback as misconceptions arise, and scaffolding instruction to address students' technical issues (Baker, 2010). Critics of direct instruction charge that the style is authoritarian and does not address student variability (Lowe & Belcher, 2012). It doesn't have to be. The UDL Guidelines remind instructors to build fluencies with graduated levels of support for practice and performance that eventually lead to independent learning. Direct instruction, which provides high levels of support, is just one tool on a continuum of teaching strategies.

Scaffolding, as a component of direct instruction, is an effective teaching method to increase student comprehension in an online learning environment. Vygotsky (1978) defined the zone of proximal development as the distance between a learner's actual developmental level and his or her potential developmental level as determined by the guidance of a more knowledgeable individual. Instructors can provide this guidance and support by scaffolding instruction.

Scaffolded instruction calls for a gradual release of responsibility to learners (Vygotsky, 1978). The first step of scaffolding is for the instructor to model the skills that students are expected to master and to explain explicitly what they are doing. This is when direct instruction is necessary. After modeling, learners can practice the skills in collaborative groups on the discussion board or by participating in collaborative projects.

Technology has made collaboration in an online course not only very possible but very productive. Google Docs is a great platform for students to use to work together in the cloud. Google Docs allows users to work simultaneously on the same document at a distance. (This book was written collaboratively in Google Docs!) It also tracks all changes so that you are able to accurately assess the individual contributions of each student in the project. The final step in scaffolding is for students to work individually to accomplish necessary learning tasks.

As you design and deliver your course, you must remember that as an instructor, you want to offer students an environment that allows them to be "buoyant" in your learning environment. In Franklin's day, swimming was thought to be a hazardous activity because of the danger of water. In our context, online learning environments can be hazardous to some learners because we lack the power to help them be truly successful. Having a solid understanding of instructor presence and its importance in all aspects of learning design will give instructors the needed confidence to create an online persona—a "humanness" that will give students the confidence they need to persist in the course and keep swimming.

@PoorRichardUDL · Testing the waters is not so bad. #pluripotentiality

REFLECTION

- What specific strategies can instructors use to build both social and cognitive presence in an online course and why are these strategies important?

- To keep all our students afloat in an online environment, we have to be "present." Consider the UDL Guidelines and the content of this chapter. How will you shift your practice to increase your online presence? Be specific about which Guidelines you will use to develop an online persona that each student can connect to, regardless of the unique characteristics of each of them.

OPTIONS FOR EXPRESSION

- Design a discussion board using the format of the Paideia Seminar. Be sure to guide appropriate goal-setting before the discussion and ask students to reflect on and assess their own participation at the completion of the discussion. For more information on the Paideia Seminar, visit www.paideia.org for professional development resources and videos.

- The section on facilitating discourse identifies strategies to help students scaffold their discussions on the discussion board. Create a handout, online video, blog, or multimedia presentation that defines these strategies for your students to use as a resource throughout the course.

5

Delivering the Package

THIS CHAPTER FOCUSES on the techniques necessary to successfully construct and deliver an online course or virtual snow day that is universally designed. These strategies are supported with research to make instruction more accessible to faculty who may not have a background in education.

We established the importance of cultivating the instructor's presence. Cultivating student presence is crucial, as well, to give learners a voice in their education. As an instructor at the University of South Florida writes: "I integrate learners' decision-making with course structure and content, while providing multiple opportunities for learners to evaluate my teaching and the course materials, enhancing their commitment and role as contributors" (Blinne, 2013, p. 41). This is why it's good practice to frequently gather data from students regarding their learning experiences so you can facilitate a course that meets all students' needs.

In *UDL Now*, Novak (2014) argues that students should be given the opportunity to assess the instructor and the effectiveness of the course throughout the course, as well as at the completion of the course. It's important to do both because they serve different purposes. Evaluations of courses are common practice in online education, but these are frequently given at the completion of a course. Any changes in practice, therefore, will impact not current students but future students. That's why it's good practice to provide frequent opportunities for students to provide feedback on your progress as an instructor so you may shift your practice to meet the needs of your current students and improve your teaching strategies. End-of-course evaluations are valuable as well, since the reflective feedback from students will allow you to make revisions for the next group of learners.

The provosts at a large New England technical college discussed the importance of this reflective feedback and decided to complete a qualitative analysis on all course evaluations to answer the question, "What advice do students have for faculty when designing and delivering a course?" The data from these student evaluations were used to provide context for, deepen understanding of, and provide insights about the communication of expectations, interaction with students, and the instructional methods of teachers.

The learners at this institution know what they want, and their requests all align to the UDL Guidelines. Arguably, many of these students have no knowledge of the UDL Framework, yet in reading their comments, provosts noted that they inherently needed all three networks of the brain to be activated in order to have a successful learning experience. If the research on UDL isn't enough, just ask your students for their recommendations and you will note the similarities. Even young students will have valuable information to share about their own preferences for teaching and learning, and shifting practice to meet their needs will improve the teaching of K–12 educators, as well.

Look at Table 5-1. In the left column, you see the themes that emerged from reading students' suggestions, and in the right column, you see the corresponding UDL Guidelines. Following the table, there will be a section for each theme, which aligns the collective student suggestion to research on best practices and delves more deeply into the UDL Guidelines to help instructors implement the strategies in an online environment.

TABLE 5-1 Student-Identified Themes for Delivery of Instruction

THEMES FOR INSTRUCTOR DELIVERY	UDL GUIDELINES
Be enthusiastic about your course.	Promote expectations and beliefs that optimize motivation (9.1 Engagement)
Follow the syllabus: Never change the course outcomes, and be realistic about what you can cover in a class, in a week, and in 10 weeks. You can't cover everything in detail...don't try! Choose the content that is most important.	• Guide appropriate goal-setting (6.1 Expression) • Support planning and strategy development (6.2 Expression) • Facilitate managing information and resources (6.3 Expression) • Enhance capacity for monitoring progress (6.4 Expression)
Give and get feedback: Return student work quickly and with appropriate feedback and grades. Ask students for their feedback, as well, to see if what you are doing is helping them.	• Optimize relevance, value, and authenticity (7.2 Engagement)
Use the readings: If there is a book or text on your syllabus, use it and have your students use it. Students hate buying a book that they don't use or don't use a lot.	• Optimize relevance, value, and authenticity (7.2 Engagement) • Maximize transfer and generalization (3.4 Representation)
Real-world application: Choose content and assignments for your course that are real-world based and significant. Students hate busy work that they cannot relate to their future careers.	• Promote expectations and beliefs that optimize motivation (9.1 Engagement) • Optimize relevance, value, and authenticity (7.2 Engagement) • Guide information processing, visualization, and manipulation (3.3 Representation)
Be fair in the way you treat all of your students. Students see when you treat someone differently from the rest. Keep high standards for all students and support them so they can achieve the standards.	• Minimize threats and distractions (7.3 Expression)

TABLE 5-1 Student-Identified Themes for Delivery of Instruction   *CONTINUED*

THEMES FOR INSTRUCTOR DELIVERY	UDL GUIDELINES
Be available to assist students with their concerns. A few minutes with a student goes a long way in preventing or solving problems. Also, read that email daily! Millennial students don't like to wait!	• Promote expectations and beliefs that optimize motivation (9.1 Engagement) • Facilitate personal coping skills and strategies (9.2 Engagement)
Provide options for graded work (assignments, quizzes, homework, labs, projects, and so on). Lots of options also provides multiple chances for students to show what they know and recover from poor grades on an individual assignment.	• Activate or supply background knowledge (3.1 Representation) • Offer ways of customizing the display of information (1.1 Representation)

BE ENTHUSIASTIC!

Okay, this may go without saying, but if you're not excited about your course, it is unlikely your students will be. Research has discovered a relationship between instructor enthusiasm and students' motivation to learn. Key to this relationship is having instructors who use their own enthusiasm to unlock the "dormant energy" inside their students so the students in turn become motivated, enthusiastic learners (Alsharif & Yongyue, 2014). To help unlock this energy, students have to perceive that the instructor is enthusiastic about the subject matter. Mastery of the subject, in combination with mastery of teaching strategies, is perceived positively by students and considered to be characteristic of the most enthusiastic teachers (Alsharif & Yongyue, 2014). In order to be an enthusiastic instructor, therefore, you have to be the master of your domain—both your content and the structure of your course. When you are able to share this enthusiasm with students, you increase their motivation to learn. Enthusiasm also relates to instructor presence; research suggests that distance learners perceive instructors to be more or less enthusiastic based on vocal delivery, apparent emotion, and level of energy, all of which are provided to students via video and audio (Alsharif & Yongyue, 2014). Our experience has shown that enthusiasm is also conveyed via the tone and style of language used in text messages, emails, and discussion boards.

Regardless of the mode of delivery, this enthusiasm needs to be communicated to students and used as a foundation to developing relationships with them. Research has found that in most courses, the discussion board is the only tool used to create interaction, and the only place where instructors build relationships with students. Since the discussion board is generally focused on peer-to-peer connection, the instructor's enthusiasm may get lost in the discussion board (Jaggars, et al., 2013). When courses are developed in this way, instructors are flat characters who become almost invisible to students.

> **@PoorRichardUDL** · Nothing happens without enthusiasm. #ignitestudententhusiasm

FOLLOW THE SYLLABUS

One of the UDL Guidelines reminds instructors to "Provide options for executive functions" for students. A universally designed syllabus or detailed lesson plan is a tool to guide students throughout the course so they can plan their time, manage all the course resources, and monitor their progress throughout the course. This document, therefore, should not change throughout the course unless it is something that the students need. If changes are made, it is recommended that you contact students using multiple mediums (course announcement, emails, video lecture, etc.) with a rationale for the change and also upload a revised syllabus as soon as possible, so students can refine their strategy for completing the course. We know it's common practice for syllabi to include a caveat that the course may change, but for students who require the document to help them succeed in the course, frequent changes become a barrier to their success. Instructors often complain that many of their students fail to read the syllabus, remember it, or refer back to it during the course. This may or may not be true, but regardless of what some students do, a well-constructed syllabus is a necessary document that is the foundation of a successful course. Syllabus content, construction, and delivery is often perceived by students to be something other than the necessary document that it is intended to be, which is why it is so important to construct it using UDL strategies and refer to it consistently throughout the course. If you scaffold the importance of the syllabus for students, they will see it as the resource it's meant to be, capable of guiding them as they monitor their progress throughout the course. When this is done consistently, changing the syllabus becomes a barrier to student success.

@PoorRichardUDL · Like any map, a syllabus gets you where you want to go. #employthysyllabuswell

GIVE AND GET FEEDBACK

So, why would an instructor change a syllabus or a lesson plan? It may be because the current structure of the course is not meeting the needs of the students. Elementary and secondary teachers who see their students every day usually get immediate feedback on the effectiveness of a lesson and respond quickly if the plan is not working. One college instructor employs a method she calls the HELP protocol, where she asks her learners four questions at the beginning of the course, which may impact the structure of the syllabus (Blinne, 2013). We have adapted these questions slightly to be applicable to online learning.

- What topics or areas are of greatest interest to us as a class? What are your goals, expectations, and learning needs for this course?

- Learning spaces can help or hinder community-building and collaboration, so how can we best adapt the virtual space to be conducive to cultivating enhanced communication?

- How can we best connect our readings and discussions to our everyday lives? Activities might include art activities, group work, journaling, YouTube videos, music, games, poetry (reading, writing, or performing), or similar.

- How can we best support and engage with multiple learning options throughout the semester (e.g., group work, presentations, films, discussions, and so on)?

If you ask these four questions at the beginning of the course and need to make changes to your syllabus as a result, you can provide students with a rationale that connects to them as learners. You can also continue to ask these questions throughout the course, as Blinne (2013) notes: "In my classroom, I return to our guiding questions repeatedly to negotiate and balance our learning needs, while still leaving room to embrace or depart from our syllabus" (p. 43).

You also need to give feedback to the students on the work that they submit to you for the assignments that you assign. Students do not like to wait for this feedback, and the more they wait the less helpful it will be. This feedback should consist of more than just a grade and include guidance and information on how the students can improve the next time. It seems like a common courtesy to our students to give them prompt and useful feedback on their work since we often give deadlines and due dates for our students to submit their work. We should return the favor.

@PoorRichardUDL • Feedback is a two-way street.
#yougetwhatyougive

USE THE READING

Most, if not all, courses generally require students to read content. Unfortunately, research tells us that many students skip required reading, since they don't believe it is meaningful or relevant to them or to their success in the course. In a large study of more than 300 students at two Midwestern universities, researchers surveyed graduate students to determine whether they felt they could succeed in the course without completing any of the assigned reading. A large percentage (31.6%) believed they could obtain an A in the class without doing any assigned readings, while 32.2% thought they could receive a B and nearly 9 out of 10 (89.1%) expected to receive a C or better (Baier et al., 2011). Sadly, only 2% of students believed they would fail the course if they didn't complete any readings.

What can we do about these findings? When we assign a text, we must make clear to students that we expect them to apply the knowledge gleaned from the text and transfer that knowledge to course assignments that are meaningful to them. The Common Core State Standards (2015), a set of college- and career-readiness benchmarks adopted in many states in America, calls for students in grades preK–12 to learn to cite textual evidence when writing narrative, informative, and argumentative texts. All students should be able to use evidence from texts to "present careful analyses, well-defended claims, and clear information" (Common Core State Standards Initiative, 2015) as a prerequisite for college and career.

College courses, including all online courses, need to build on the Common Core's expectations by requiring all students to read complex texts and respond to ideas with specific evidence from these texts. Assignments that require this high-level skill from students will ensure that students will see the text as an integral part of the course. Also, if students do not read the text and reflect on those complex ideas, they will not be successful. If the reading is assigned, it needs to be meaningful.

@PoorRichardUDL • Readings are often ignored because the teacher gives the info in class! #haveyouconsideredflipping

REAL-WORLD APPLICATION

Students hate busy work. That should come as no surprise: We professionals hate busy work, too. If we're going to invest our time in a project, we want it to be worth that time. Not only will real-world applications increase student engagement, but these assignments increase content knowledge and maximize students' ability to transfer knowledge to their chosen careers.

Numerous research studies have focused on real-world application of content in K–12 settings and at the college level. In *Powerful Learning: What We Know About Teaching for Understanding*, Darling-Hammond (2008) referenced a nine-week project where fourth- and fifth-grade students worked collaboratively to define solutions for housing shortages in countries across the world. The project-learning students scored significantly higher in both critical thinking and problem solving at the completion of the unit than the control group. In a higher education study, the authors wanted to determine if real-world application of statistics would result in the spontaneous transfer of statistical knowledge, as compared to a control group who received more traditional assessments (Daniel & Braasch, 2013). Results indicated that students who received the real-world application exercises had greater statistical knowledge at the end of the semester as compared to the control students. Even though these students received the same instruction, the engagement principle was emphasized more with the students who were able to apply their knowledge in authentic, meaningful tasks.

At Baylor University, the school of business operates on a "simple and valuable principle," which is, "why not equip students with sales skills and real-world experience before they enter these businesses as professionals?" (Krell, 2010, p. 8). This guiding principle is at the core of UDL. If we want students to transfer the knowledge they learn to their chosen careers, we need to scaffold that process by designing activities, assignments, and assessments that allow them to apply new content in the same way they will have to when they graduate. This allows students to experience firsthand how their emerging knowledge will transfer to a successful future.

@PoorRichardUDL • Real life = real education. #makeitreal

BE FAIR! AND THAT MEANS TRANSPARENT!

For nearly 50 years, researchers have examined the correlation between student grades and instructor evaluations. Early research focused on student achievement and its relationship to instructor quality. In the 1960s, Adams (cited in Tata, 1999) examined whether students negatively evaluate their instructor when they are graded poorly in a course. He found that there is not always a correlation between low grades and negative evaluations. Rather, negative evaluations are more often tied to perceptions that the grading was *unfair*. This connects to research on procedural justice: that is, whether instructors follow course procedures consistently. For example, if an instructor allows some but not all students to use graphic organizers or to hand in work late, all students are more likely to evaluate the instructor as being unfair. Students expect instructors to apply grading procedures fairly to all students. Therefore, when students believe that grading procedures are biased, they are more likely to evaluate the instructor negatively (Tata, 1999).

We would add that an important part of fairness is transparency. A fundamental UDL principle is that goals should be clear—and referenced frequently as part of an ongoing system of progress monitoring. That way, learners always know "where they stand" in a lesson, unit, or course; they know the point of what they're working on (Meyer, Rose, & Gordon, 2014). When instructors heighten the salience of goals and objectives, support students as they monitor their progress, and encourage students to self-assess their work, students will have a stronger partnership with the instructor and will have a clearer understanding of what instructors expect.

Our friend Ben Franklin also has said, "Do not fear mistakes. You will know failure. Continue to reach out," and even though the concept of educational testing or self-assessment was not in fashion during his time, his comments on "not fearing mistakes" really fits well with the concept of self-assessment. Self-assessment is a great opportunity for our students to see how they are doing without all of the issues associated with in-class testing or assessment: mild anxiety at one extreme and fear and loathing at the other. UDL favors empowering students to learn a concept in the way that works best for them. Self-assessment allows students to see if they are learning without the risks of a test for a grade. Providing self-assessment options will also help students prepare for the graded assessments that are eventually a part of all instruction. Self-assessments can also be provided in a wide variety of modes of engagement.

Therefore, if instructors align grading procedures to the UDL Guidelines, bias is proactively minimized, since students have multiple opportunities to receive mastery-oriented feedback, and they are consistently encouraged to self-assess their work and reflect on their learning, in partnership with instructors.

@PoorRichardUDL • Being transparent makes learning more successful. #keepitclear

BE AVAILABLE

Instructor availability is not only a student suggestion for increased satisfaction, but research tells us that it improves their outcomes significantly. One study compared success rates in 137 online courses with a total of 2,432 students (Orso & Doolittle, 2012) and found that 59% of students were successful in online courses. However, the success rate of students in online courses with "outstanding instructors" had a success rate of 82%. The authors defined an outstanding instructor as one who responded at least three times daily to all online course emails, graded all papers within 48 hours of submission, offered specific feedback on all written work, and were responsive to students' needs (Orso & Doolittle, 2012).

Being available and responsive, in this study, had a significantly positive effect on student success. When the authors shared this study, however, they noted the displeasure of many faculty members who felt as though this level of availability was "coddling" the students. Through the UDL lens, this support is not coddling, but rather using techniques that minimize barriers that prevent students from succeeding in a course.

This is all the more important in an online course because instructors don't have the opportunity to acknowledge students and answer questions as easily as in a face-to-face course. For example, in a traditional classroom, instructors are able to acknowledge all students in every class by making eye contact, nodding, and responding to questions and comments as the class moves along (Palenque & DeCosta, 2014).

PROVIDE OPTIONS

Providing options is a critical component of the UDL Framework. The nine UDL Guidelines remind us to "provide options" for perception; language, mathematical expressions, and symbols; comprehension; physical action; expression and communication; executive functions; recruiting interest, effort, and persistence; and self-regulation. Throughout this text, we have provided suggestions for the types of options you can design for students, so at this point, we feel it would be useful to revisit our hypothetical students to see how these various options would minimize barriers to their learning.

Coco needs options to decrease her anxiety. She needs to know if it's okay if she audio-records or hand-writes the assignment, scans it, and then posts it on the LMS. In class, she is always at the teacher's desk. Since the teacher may not be available 24/7 in the LMS, directions need to be available and clear so Coco knows, without a doubt, what she is allowed to do. Also, the grading criteria will need to be clear; otherwise, she will work herself into a panic thinking that she will fail the assignment.

I would also be important for Coco to know that if she has a question during the virtual snow day, she will be able to ask the question on a discussion board and get a prompt answer.

Kriti may need all the representational options at one time or another, but the instructor's enthusiasm, fairness, and availability will need to be extremely evident for Kriti to succeed. Kriti's self-confidence will have to be supported often, so she will also benefit from a variety of grading options that she can use to find the assignment that "feels right" for each topic.

June is a rule follower. Having a syllabus that is a true representation of the course requirements will be a treasure for her to print, follow, check off, and look ahead. It will provide a connection between her "traditional sense of a course" and the virtual course that she is participating in online. Finally, having an instructor who is available to answer questions and virtually hold her hand through early problems of connectivity and computer questions will prevent June from getting lost in the course.

José might not need any instructor support. A well-planned and complete syllabus will give him everything he thinks he needs to succeed in the course. He will complete assigned reading for the content. He will do the real-world assignments for the experience. He will appropriately challenge you as the instructor so he can improve, and he will take the feedback you give him and use it to improve his next assignment.

Ray is going to need the flexibility of grading options so that he can tailor what he needs to do for the course within the time that he has available. He is also going to need the real-world applications to keep him motivated and engaged on a weekly basis. Ray can do the work, and wants to do the work, but his job and family will pull on him as well, so he will need lots of feedback to reinforce the progress he is making in the course.

It's remarkable that a group of students who likely have no background in instructional design and UDL are able to summarize the skills necessary to increase student outcomes in learning environments. Don't only take the advice of these students, however. We encourage you to collect feedback from your own students, throughout your course, regardless of their age, and make adjustments so barriers in your course design are minimized.

@PoorRichardUDL · Student feedback = educational gold. #listentothestudentstoo

In the next section, we will bring all of these best practices together as we discuss how to build and design your discussion board so you can deliver instruction that meets the needs of all your students.

DISCUSSION BOARD

Discussion boards are a staple in asynchronous online learning environments, and a great place for students to connect, reflect on their learning, and participate in academic discourse that is a hallmark of face-to-face classes. In order to maximize the benefits of discussion boards and promote student critical thinking, however, instructors need to channel their inner UDL.

Discussion boards must be intentionally built to foster student collaboration and critical thinking. When they aren't designed that way, online discussions can result in boredom, inactivity, low-quality responses, and responses that lack depth (Hall, 2015). Simply participating in a discussion may not result in student learning, which is why instructors must moderate discussions so students are prompted to follow the five-step process necessary for critical thinking: identifying a problem, exploring the problem, suggesting a solution, judging the solution, and implementing the solution (Hall, 2015).

If we need students to practice critical thinking, that begins with the quality of the discussion prompt, which needs to help students identify a problem and explore it with the help of their classmates. As with all things UDL, however, student engagement needs to take a prominent role. Not only do prompts need to result in critical thinking, but they also have to be differentiated, meaningful, and relevant to student lives.

To do this well, instructors must harness their creativity. Make every discussion different. Change the question style, vary the response requirements, or invite a guest moderator or speaker. Research by Hemphill and Hemphill (as cited in Hall, 2015) has found that inviting a guest speaker into the discussion board results in half of the students responding using upper-level critical thinking in the first discussion and 70% of students using upper-level critical thinking in the second discussion. This is only one way to change the pace of the board, but there are others. For example, you could have students respond to videos, artwork, presentations, and so on. Also, as you build your discussion board for the term, use the sequence of the discussions to scaffold the learning of the mega-objective of the course.

Also remember to develop each discussion using the techniques in the Paideia Seminar, discussed in Chapter 4. Summarizing key concepts and goals before, during, and after the discussion increases instructor presence and helps students to monitor their own progress, reflect, and assess their own learning, which helps to build fairness when you are assessing their responses, as we mentioned previously.

To recap, you want to make sure you have a question that will allow students to explore a problem and build critical thinking. Also, make sure that you require your students to use the readings and other content from the course in the discussion posts so you are heightening the importance of the class texts. This should be more than just the listing of facts and ideas from the reading. Ask the students to use the information from the readings in the support of their answers to the discussion question. This is exactly what you would do in class face-to-face, but it is both more important to do this in an online class (since you can't have a face-to-face discussion) and easier to assess the quality of student understanding (it provides a record, and the student will have to defend his or her point of view argument). Next, try to phrase the question in a personal way to build engagement: "What does this mean to you?" or "How have you dealt with this situation in the past?" Finally, always provide guidance on proper course "netiquette," This seems to be less and less of a problem, but it should never be forgotten.

Once you have the right discussions set up, create clear rules of engagement:

- Describe the types of responses required—will there be a primary post and then a follow-up post to dig deeper into the ideas that are presented?

- If you wish your course to truly be UDL, you should be prepared to allow and encourage multiple forms of engagement and expression. It doesn't (and shouldn't) always have to be just text. Free services like YouTube or Screencast-O-Matic allow for a quick aand easy way to record video (if you have a webcam) or screen-capture movies. Smartphones provide an even easier option.

- Will students be able to see other posts before they make their first post? Many LMSes have a feature that hides the first posts until after a student makes his or her first post. This encourages more unique first posts that are unaffected by fellow students' posts.

- How much will you, as the instructor, participate in the discussions? You need to monitor and moderate the discussions, but that does not mean you have to participate in each conversation, although you definitely want a plan before delivering the course (Mazzolini & Maddison, 2007). Our personal preference is to participate often in the first few discussions so that our students know that we are present. Our participation is usually limited to answering direct questions, cleaning up any misdirection or misinformation, and encouraging students to add additional information or consider another angle. After the first few discussions, we continue to watch, but reduce our participation to asking more questions so we are just moderators, helping our students build critical thinking skills.

- How will you provide mastery-oriented feedback to the work in the forums? You will need to provide feedback to your students and you will have to give prompt feedback. Will it be individual feedback or group feedback? Will it be in the form of text, video, or something else? As always, you will want to listen to your students and respond to their concerns.

- How will you evaluate participation in the discussions? Will the discussions be graded? Will you have minimum standards for posts and post lengths? Will there be a rubric? We believe that an individual grade is important, in order to instruct each student in the proper methods of participation. Obviously, if the discussions will be graded, that information needs to be placed front and center in the discussion board description.

- You need to make clear the due date(s) for each of the components—will the first posts be due on Thursday evening by end of day? Will the follow-up responses be due by Sunday end of day? Whatever you set up for due dates, be consistent and fair when applying them. You need to have procedural justice so students do not think you are "unfair." A perception of unfairness could act as a barrier for some students.

During the first week of the course, have students introduce themselves in the first discussion, and state their goals for taking the course and their long-term goals. Require the other students to participate in a conversation based on their posts. This usually happens naturally as students notice similarities, disciplines, or backgrounds. In a program, students will also have taken courses together and will be glad to "see" each other again.

You also need to remember that sometimes, in these discussion boards, students need to "see you" in online course videos. As we mentioned in the creation of your course syllabus, you have to make this course yours by letting your personality show in your writing style and how you interact with your students. Our personal preference for doing this is to use video often to convey messages, describe lessons, and give feedback. Our videos are very targeted and they are not mini Hollywood epics. They are watchable videos that convey a short message without major technical distractions. We believe that the videos convey our personality, provide important information, and reduce the distance between the students and the instructor. We are not big believers in using video to lecture, but video lectures do have a place in the discussion board if they are constructed with online realities in mind. That reality is an acceptance that watching a video of a lecture is not the same as being in a face-to-face classroom. So, if you wish to create lecture videos, keep them short; 5–10 minutes maximum. According to research at EdX, the optimum length for an online video is 6 minutes (Guo, 2013).

Chunk the content into small segments that you surround with activities. Plan to have the students do the work that is necessary to successfully complete the course—active learning, less lecture. A short video is a great way to introduce a topic, give the necessary background before a discussion, or provide students with mastery-oriented feedback.

Set up a separate discussion board as an ongoing Q&A for questions that come up from time to time. Encourage students to feel free to answer questions from their peers if they can. Depending upon the LMS you use, the system could send you an email whenever someone posts to the board.

If you were teaching in a face-to-face course you would be able to assess the participation level of each student in your course just by observing them. In an online course, the only way that you can do this is by monitoring course participation online and contacting nonparticipants. One way to help do this is to only open the course materials weekly to maintain group cohesion or enforce a sequential path through the course. This will minimize the amount of student work that you will have to track each week, especially if you set up a policy for late posts and work that minimizes the amount of late work you have to manage.

When speaking of self-assessment, we shared a quote from our friend Ben Franklin. "Do not fear mistakes. You will know failure. Continue to reach out." We encourage you to take this advice as you build or revise your online course. As your knowledge of UDL continues to evolve, and as you work in partnership with your students to minimize barriers to their learning, you will learn a little about failure, but this failure is not permanent or complete. Use the UDL Guidelines and you will be empowered to make adjustments in your learning cloud to increase student learning outcomes.

REFLECTION

- When you deliver an online course, flip your class, or deliver a virtual snow day, it's important to use instructional strategies that eliminate barriers. Compare the instructional strategies identified in Table 5-1 with those in your own practice. How are your instructional strategies similar? How are they different?

- A research study noted that an outstanding instructor was one who responded at least three times daily to all online course emails, graded all papers within 48 hours of submission, and offered specific feedback on all written work (Orso & Doolittle, 2012). Are these expectations reasonable? Why or why not? Respond to this assertion in a blog, online video, a spirited chat with a colleague, or in an online discussion with your students.

OPTIONS FOR EXPRESSION

- It's important to have clear rules of engagement for your discussion board. Examine your expectations for your discussion board and then cross-reference each item with the UDL Guidelines. If your expectation does not align to UDL checkpoints, identify possible barriers, and eliminate or revise the expectations so all students have increased opportunities for access and success.

- Share your discussion board expectations with students and ask for their feedback about the appropriateness of each requirement. Ask them to comment on each item's fairness, rigor, applicability to the real world, and value. This could start a meaningful conversation about the expectations for discussion and it may impact how you design discussions in the future.

- Review any video lectures you have posted in your courses. How long is each video? Remember that the ideal length is 6 minutes. After your review, consider removing any videos that are longer than 6 minutes or break up the video into small chunks with opportunities for discussion and reflection between each section. If you don't have any videos, record your first and post it online. Your students will appreciate the opportunity to get to know you.

6

Scaffolding Time Management

ONE BARRIER that is particularly relevant in the cloud is time management. Many learning experiences are asynchronous, thereby making attendance at a set time less of an obligation. This chapter will explore how online instructors can use the expression and engagement Guidelines to scaffold the process of time management so students can be successful.

Good ol' Ben knew what time was all about, too. He not only invented the concept of daylight saving time but he coined the expression "Lost time is never found again." Time is precious, and life in the 21st century seems to consume time so fast that it often feels like it is gone before we get a chance to do anything with it. Life is busy, and adding major events to the mix like completing an online course, taking a virtual snow day, or getting a college degree can make it seem like there is just not enough time in the day.

Not to dive too deep into physics, but the notion of time is actually an illusion. As Callender (2010), a philosophy professor at the University of California, San Diego, tells us, "The present moment feels special. It is real. However much you may remember the past or anticipate the future, you live in the present. Of course, the moment during which you read that sentence is no longer happening. This one is." Time is all about anticipating what needs to get done in the future so that in each present moment the most important tasks are being completed before they become things of the past.

Research on time management tells us that there are a number of variables that make task initiation, and therefore task completion, much more likely. These variables align to the UDL Guidelines and allow us to scaffold time management so students can be successful in our online learning environments.

TABLE 6-1 Variables That Affect Time Management and Task Completion

VARIABLE	RESEARCH (TU & SOMAN, 2014)	UDL GUIDELINES
Time perception: Provide students with a visual representation of the time necessary to complete each task (e.g., timeline, demonstration, calendar)	Visual cues that facilitate categorization of time could increase task completion.	• Offer ways of customizing the display of information • Illustrate through multiple media • Guide information processing, visualization, and manipulation
Goal striving: Have students set goals for themselves regarding task completion	Merely prompting consumers to categorize task deadlines will increase task initiation and completion.	• Guide appropriate goal-setting • Support planning and strategy development
Choice architecture: provide choices that will allow all students to be successful	Design choice environments that nudge consumers into undertaking a desired course of action.	• Optimize individual choice and autonomy • Facilitate personal coping skills and strategies

How can we, as instructors, leverage time perception, goal striving, and choice architecture to meet the needs of our students? We must rely on UDL.

@PoorRichardUDL · Lost time is never found again. #benstrikesagain

TIME PERCEPTION

Having a syllabus that lists assignment due dates is important so students can visualize when tasks need to be completed, but often, this list of deadlines is not presented using multiple means of representation. For example, imagine the following are the assignment deadlines for the first three weeks of a course:

> ▶ **Complete the assignments by the following dates:**
>
> * October 10: Review course syllabus, review all material in the Week 1 course module, and take short quiz #1
>
> * October 12: Read Chapters 1 and 2
>
> * October 14: Reply on Week 2 discussion board
>
> * October 21: Read Chapter 3 and post response #1 to Week 3 discussion board
>
> This view provides only one visual representation of the assignment due dates. Sharing the same information in a table may help students to categorize their time more effectively (see below).

WEEK	ACTIVITY/ASSIGNMENT	DUE DATES
1 Course begins October 1–7	Short quiz #1	October 10
2 October 8–14	Read chapters 1 and 2 in text. Post to Week 2 discussion board.	Original post: October 12 Replies: October 14
3 October 15–21	Read Chapter 3 and post response #1 to discussion board	Response #1: October 21

In alignment with the UDL Guidelines, we also recommend illustrating due dates using multimedia. A simple way to do this is to load all assignments into the LMS calendar. Students can then customize the calendar to view assignments by the date, week, or month.

Although having multiple visual representations is important, students may not have an appropriate understanding of how much time they should set aside for each assignment, and therefore, the table view and calendar view also have limitations.

Generally, when we look at a calendar, we look for things that we should be doing today. Like, right now. We need to encourage students to look ahead and plan out their time so they are more likely to complete tasks by the deadlines. This will involve students setting goals for themselves, but this will be difficult without background information on how long they should set aside for each task.

We can facilitate time-management strategies by providing students with an estimation of how long each assignment may take, so the goal-striving process will be easier to complete.

The amount of time that it will take each student to learn is variable. However, we continue to assume that all students can learn within a defined amount of time—5 minutes, 30 minutes, 2 hours, 1 day, 1 week—and we build our courses within those assumptions and create courses that follow a standard quarter or semester. Certainly, secondary and postsecondary institutions segment more complicated content into a variety of sequential courses (BIO I, BIO II, etc.) but the time frame is "king." This is why it's important that students understand the minimum amount of time they should set aside for each week to complete necessary tasks. A great way to share this is to ensure that all students have the necessary background information regarding the Carnegie Unit.

@PoorRichardUDL • Different people perceive time differently. #whatdoesoneminutemeantoyou

The Carnegie Credit Unit

The Carnegie Credit Unit, a staple in education, requires instructors to design and deliver learning opportunities that meet minimum time requirements. A recent report from the Carnegie Foundation (Silva, White, & Toch, 2015) examined the role, function, and uses of the Carnegie Unit in educational institutions from elementary school to graduate school. After this 2-year study, the authors report that the Carnegie Unit "continues to provide a valuable opportunity-to-learn standard for students in both higher education and K–12 education, where inequitable resources and variable quality are more the rule than the exception" (p. 5).

Opposers of the Carnegie Unit argue that instructors should be focused on what students actually need to learn, and not the time that they should spend on the task. To support this argument, opposition cites the policy of many colleges not to accept transfer credits, regardless of the grade the students received, as evidence that many institutions do not equate time with learning.

WHAT IS A CARNEGIE UNIT? (FROM SILVA ET AL., 2015, P. 8)	
In high school	The standard Carnegie Unit is defined as 120 hours of contact time with an instructor, which translates into one hour of instruction on a particular subject per day, 5 days a week, for 24 weeks annually. Most public high schools award credit based on this 120-hour standard (one credit for a course that lasts all year, or half a credit for a semester course). And, while state and district coursework requirements for graduation vary, most states require a minimum number of units, typically expressed as "Carnegie Units." A typical high school student earns six to seven credits per year over a 4-year program of high school.
In higher education	In higher education, students receive "credit hours," a metric derived from the Carnegie Unit and based on the number of "contact hours" students spend in class per week in a given semester. A typical three-credit course, for example, meets for 3 hours per week over a 15-week semester. A student, then, might earn 15 credit hours per semester (15 is standard full-time registration for a semester, 30 for an academic year) en route to a 4-year bachelor's degree requiring a total of 120 credits.

Since the time required for students to complete tasks varies greatly, time management must be individualized for students, which is why they need to create their own time-management goals within the confines of the Carnegie Unit, because it's not going away any time soon.

In the Carnegie Report, the authors note that educational institutions of learning face barriers when it comes to eliminating the Carnegie Unit and moving toward proficiency-based learning. In public K–12 education, the most notable barrier is public perception. The authors note that, "a Carnegie Foundation analysis of state policies for this report found few prohibitions against school systems uncoupling course credits from instructional time" (Silva, White, & Toch, 2015, p. 24). Therefore, K–12 institutions who are interested in providing a more flexible model of instructional time should explore options afforded in their respective states. Virtual courses are more likely to offer competency-based instruction, as opposed to time-based. Since tracking seat time is difficult in a virtual classroom, instructors assess multiple measures throughout the course to determine if students meet proficiency requirements (Archambault, Kennedy, & Bender, 2013). Students, therefore, can complete work quickly and not have to log additional time in their seats to meet requirements.

In higher education, it comes down to money, specifically, "the federal government's requirement that most students taking part in the $150 billion federal financial aid program attend colleges or universities using Carnegie Units" (Silva, White, & Toch, 2015, p. 25). Considering that finances are a barrier for many adult students, we must operate in units of time when delivering instruction in higher education, because if we eliminate the barrier of time, we put up a financial barrier.

So what does that flexibility look like when operating in the Carnegie Unit? That depends on the students, their schedules, and the goals they have for their task completion, which leads us back to the second strategy for scaffolding time management: goal-striving.

GOAL-STRIVING

Once we have our course deadlines, it's helpful to remind students that they should plan to spend a minimum of 3 hours a week per credit on task completion. Again, in a virtual environment, this will vary, but it's important to have students carve out time to complete assignments. As a first step, ask students to look at their weekly schedule and figure out where they can carve out those 3 hours per credit for each online course.

Let's do the math for José, a full-time online college student who works full time. In a standard 18-credit hour quarter we are expecting 3 hours of academic engagement for each credit hour, or 54 hours ($3 \times 18 = 54$). There are 24 hours to a day and 7 days a week for a total of 168 hours of possible time for students to live, work, and study.

ACTIVITY	PER DAY	PER WEEK
Sleep	8	56
Eat	2	14
Commute	1	5
Work	8	40
Class/lab	3.6	18
Homework	7.2	36
Totals	29.8 (-5.8)	198.8 (-30.8)

As you can see above, there are simply not enough hours in the day, and we haven't even factored in time at the gym, time with family, or any enjoyable hobbies or activities. Is it any wonder that students have difficulty managing their time? It would be a lot easier if there were more of it for them to manage. Alas, there is not, so we need to help José come up with a strategy so he can complete all of his assignments.

As a first assignment, therefore, you could ask students to reflect on their calendars and create a plan for completion of work. This would even be helpful for our student Coco, on virtual snow days, since she will be worried that she will not have enough time to access learning on the computer. In class before the first snow day, her teacher could encourage her to wake at 6:00 am, log on the computer, print out the to-do list and any readings, and then eat breakfast before sitting down at 7:30 to begin working.

@PoorRichardUDL · Time is an herb that cures all diseases. #aphorismskeeponcoming

Once students have a plan, they are more likely to complete the task. They are all the more likely to complete it if it's something they are interested in.

CHOICE ARCHITECTURE

Time is perception. When students are engaged in a task, they are more likely to enter a time warp, or what Mihaly Csikszentmihalyi calls *flow*. In an interview with Sobel (1995), Csikszentmihalyi explains flow as "a continuum, a combination of dimensions of experience, beginning with a challenging activity requiring skills with clear goals and feedback. The person becomes utterly absorbed in the activity, concentrating so intently he or she drops all self-consciousness and loses the sense of time." Clearly, time is easier to manage when you're not concerned with clocking in and out to complete an assignment.

Based on the concept of flow, learners lose sense of time when they are truly engaged. Providing choices makes it more likely that learners will choose an activity that allows them to operate in this time warp. We all have examples in our own lives, when we were caught in flow while learning. We'd like to share our own flow experiences and ask you to consider yours. We imagine that many, if not all of you, will find that you were able to transcend into a state of flow because you were given choices that allowed you to find happiness in your learning.

KATIE NOVAK	TOM THIBODEAU
I have always been intrigued by antique furniture. I look at old dressers and imagine the people who folded their clothes and tucked them neatly in the drawers, warped and stained by time. My uncle John, a master craftsman, has always found treasures like this and refinished them.	I once was in a senior seminar course that required a substantial project presentation of an American novel of our choice (from a defined list). We had to read a book a week for this course. I chose *The Naked and the Dead* by Norman Mailer. I loved the book—all 700 pages—so much so that I reread the book over the weekend to prepare for the presentation and cataloged all of the major themes of the book. There were four or five pages of themes. I spent most of the weekend preparing. The presentation went well and the professor was impressed. I don't know why I chose to do as much work as I did for this particular project but something about the project and the process of cataloging the themes got me so involved that the time didn't matter at all.
I have spent countless hours with my nose deep in books, my eyes fixated on shows like *This Old House*, and more time than I care to admit walking up and down aisles in Home Depot studying paint thinners, hand sanders, and polyurethane as homework.	
The thing is, it's homework that makes the real work more rewarding. When I start working on a piece of furniture, I am lost. The time I set aside for grueling labor is not work at all, because when I am refinishing furniture, time does not exist.	

The point here is that it is possible for us to motivate our students to achieve independence in an academic exercise if we find all of the right "buttons" for our students, we scaffold the assignments in such a way that the students learn how to manage their time most effectively, and we don't have the students waste their time on busy work.

When Chickering and Gamson (1999) defined seven principles for good practice in education, online learning was not on their radar. However, they listed seven timeless strategies for effective teaching. They argued that effective practice encourages contact between students and faculty, develops reciprocity and cooperation among students, encourages active learning, gives prompt feedback, emphasizes time on task, communicates high expectations, and respects diverse talents and ways of learning. It's almost like an infomercial for UDL.

What is interesting about their principles is their emphasis on time on task. In order for students in any environment to be successful learners, they need to dedicate the time necessary to complete required tasks.

TIME MANAGEMENT

One great strategy for modeling time management involves breaking up larger projects into small components. Let's return to our *Teaching with Technology* course. The final project for this course includes a reflective narrative, an annotated bibliography, and three lesson plans, with development curriculum, that integrate the use of technology appropriately.

▶ Introduction

Your final project will contain representative work that you've produced this session, along with a reflective narrative on your pedagogy (either written or in video or audio form), new sample lessons that incorporate computer or network technology, and an annotated bibliography. Sample projects and assignment rubrics are posted in the LMS under "Assignments."

Please include the following items in a single point of access (for instance, in a single Microsoft Word file or Google Doc, a website, etc.).

Contents

1. *Develop a reflective narrative* on the relationship and role of computer and network technologies on your teaching philosophy and practice. This narrative should provide a relatively broad perspective and can be presented from the first-person point of view (if desired), using relatively informal language. This narrative can be written or recorded using audio or video.

2. *Produce an annotated bibliography* that contains 10 items that would be of interest to others who want to incorporate information and network technologies in teaching their specific disciplines. Please use APA format. A sample will be posted in the LMS.

3. *Draft three sample lessons* that leverage information or network technologies, along with a brief discussion of the specific outcomes associated with the technology use.

4. *Produce sample video, audio, or web content* that integrates into one of your lesson samples and uses the most appropriate technology that services your lesson. In a simple reflection, include a link to your content and discuss the design of your content (approximately 250–500 words, but it can be longer if you wish). Again, this discussion can be written or recorded using audio or video.

To ensure that students are working toward the development of their final project, we need to be very clear in our expectations and help our students manage their time by breaking up the major projects into weekly components. This eliminates any "busy work" that a student might think is a waste of time, and it requires them to begin their final project early so that it doesn't become a huge project the week before it is due. The course schedule would look something like this:

▶ Course Schedule

Week 1: Asynchronous Communication Strategies: Primary question of the week: Why do we need to communicate at all with our students?

1. Email and listservs

2. Discussion boards, blogs (with and without subscriptions)

3. Wikis

4. **Final Project:** Step 1: Topic identification

Week 2: Content Management: Primary question of the week: How do we efficiently manage online content?

1. Online document publication (standards and best practices)

2. Document archives and document access

3. Managing multiple course sections and managing multiple platforms

4. **Final Project:** Step 2: Annotated bibliography

Week 3: Synchronous Communication Strategies: Primary question of the week: Is this worth all the effort that it will take?

1. Virtual classrooms and whiteboards

---

2. Webinars, Google+, Skype, and other live multimedia communication tools

3. Optional webinar (synchronous video conference) will be scheduled this week.

4. **Final Project:** Step 3: Sample content for one of the lessons

Week 4: Social Media Interaction: Primary question of the week: How do we leverage the tools that our students use on a daily basis?

1. Communicating with Twitter, Facebook, and so on

2. Using social media tools for course instruction and collaborative interaction

3. **Final Project:** Step 4: Submit lesson one

Week 5: Multimedia Content: Primary question of the week: How can multimedia help my pedagogy?

1. Creating simple audio and multimedia content using Audacity and/or Screencast-O-Matic

2. Posting and making multimedia content accessible

3. Multimedia instructional design (standards and best practices)

4. **Final Project:** Step 5: Submit lesson two

Week 6: Writing and Research Tools and Strategies: Primary question of the week: What writing and research tools can we use in our online teaching?

1. Feedback and response strategies

2. Mind-mapping and idea-generation strategies

3. Reference and research tools

4. **Final Project:** Step 6: Submission of your final project with your third lesson and personal reflection

The weekly details in the course LMS would then look something like this:

▶ Course Schedule/Outline

Week 1 (October 1–7): Asynchronous Communication Strategies: Please see the Week 1 module for more details.

A. Primary question of the week: Why do we need to communicate at all with our students?

    1. Email and listservs

    2. Discussion boards, blogs (with and without subscriptions)

    3. Wikis

B. Reading:

- "Disrupting Ourselves: The Problem of Learning in Higher Education" by Randy Bass: http://www.educause.edu/EDUCAUSE+Review/EDUCAUSEReviewMagazineVolume47/DisruptingOurselvesTheProblemo/247690

- "How Disruptive Innovation Changes Education: Q&A" by Martha Lagace with Clayton M. Christensen, Michael B. Horn, and Curtis W. Johnson, August 18, 2008: http://hbswk.hbs.edu/item/5978.html

- "So You've Got Technology. So What?" by Richard A. DeMillo: http://chronicle.com/article/So-Youve-Got-Technology-So/131663/

- "Disruptive Education Technology: Helping Kids Learn" by Michael Horn: http://images.businessweek.com/ss/08/10/1021_education_tech/index.htm

- "Implementing the Seven Principles: Technology as Lever" by Arthur W. Chickering and Stephen C. Ehrmann: http://sphweb.bumc.bu.edu/otlt/teachingLibrary/Technology/seven_principles.pdf

---

C. Videos: These videos are an alternative to the readings above.

- "Learning, Innovation and possible futures of Georgetown" by Randy Bass: http://vimeo.com/68049901

- "Dr. Clayton Christensen discusses disruption in higher education" by Clayton M Christensen: https://www.youtube.com/watch?v=yUGn5ZdrDoU. Watch around the 50:00-minute mark for an interesting discussion about online learning and teaching individual students.

D. Activities

- Post to the discussion board—Introductions: Post a personal introduction in this forum as soon as convenient. Please see the Week 1 module for more details on how to navigate the discussion forum.

- Post to the discussion board—Instructional Technologies—A Starting Point: Initial post by Saturday, noon; three follow-up posts due no later than Tuesday, noon. Please see the Week 1 module for more details on how to navigate the discussion board.

- Email posting on Wikipedia to listserv (send assignment to COWC-TWT@yahoogroups.com): no later than Tuesday, midnight, though sooner is better. If you need help with the listserv check out this link: https://info.yahoo.com/privacy/us/yahoo/groups/details.html

- Send email to your instructors with your choice of a topic for your final project.

E. Need help?

- Tutors at the Academic Skills Center (ASC) are available to help you improve your reading ability or your writing skill. Call 202-555-0115.

- Need a text-to-speech app? Please check out these links: https://digitaltext.wordpress.com/at-tools/text-to-speech-options/free-text-to-speech-options/

- https://www.apple.com/accessibility/osx/

By scaffolding the project over multiple weeks we allow students to focus their attention on one component at a time. These components represent different tasks and different modes of expression that require different skill sets. Breaking up the tasks will help the students manage their time effectively.

FINAL THOUGHTS

As you think about time management, remember to help students learn how to manage their time as *part* of the course. Many education institutions offer time-management seminars and enrichment courses. A Google search delivers 215 million hits on the topic. However, offering a course as a supplement to time management is much different than building time-management skill development into a course or virtual snow day. Obviously, the challenge will be to bring the time-management exercises to the course without adding more time commitments to the students.

Recognize that time is an asset that you are always spending, and it can never be replenished or replaced. Additional time-management skills need to become part of the assignment so as not to take away from the already limited time that students have to complete that assignment.

Second, make real-world connections about the importance of deadlines. For example, when teaching a video production class, it would be important that students understand just how deadline driven the world of media really is. It would be beneficial for students to have a productive struggle as they race to meet deadlines, because this will be crucial in their career. At the beginning of the course, time-management skills are scaffolded, so at the end of the course, they can be told, "You are preparing to join a deadline-driven business, so if you don't submit your projects by the deadline, I don't want them."

Time management is not just about being successful in an online course or lesson; it's about being successful in life. Design your course so you explicitly teach students what they need to do to meet deadlines. Heighten the salience of your goals and deadlines, encourage students to reflect on their schedules and set goals for task completion, and provide engaging choices so they are motivated to spend their time learning. Finally, provide graduated levels of support so after the course is over, they can transfer those important skills to the real world, where deadlines are often inflexible.

> **@PoorRichardUDL** · Building skills takes time and needs to be taught. #engagemeandIlearn

REFLECTION

- Think of a learning experience when you entered a state of flow and time didn't seem to exist. Analyze this learning experience through the UDL lens. How can you create a similar experience for your students in a virtual environment using the UDL Guidelines?

OPTIONS FOR EXPRESSION

- Read more about Mihaly Csikszentmihalyi and his concept of flow. After your research, design a lesson to teach your students about the concept and to reflect on educational experiences that take them into a state of flow.

- You have 168 hours a week to live, sleep, eat, and work. Using a chart, or other representation, depict how you manage your time each week to meet your professional and personal goals. If you are a multitasker, make that clear in your depiction. Consider sharing this with your students so they have a model to manage their own schedules.

- Create an assignment for students where they have to develop a schedule for themselves to complete necessary classwork while also continuing to meet personal and professional obligations. Encourage a reflection so students can determine how their schedule will have to change to be successful in your course.

7

Application to the World of Hybrid-Online

THIS CHAPTER WILL delve into two relatives of online coursework: flipped classrooms and virtual snow days. We will discuss practical considerations for these two cloud experiences so all barriers are eliminated and all students can succeed in these environments.

In this text, we argue that variability among learners is significant, and regardless of age, learners need their instructors to increase their engagement, make learning meaningful, and provide options for representation and expression. This is true in settings that are completely online, as well as those that are hybrid-online. In this section, we'll explore two environments that differ slightly from online coursework provided in virtual K–12 schools and postsecondary institutions: flipped classrooms and the virtual snow day.

@PoorRichardUDL • Online learning can happen at all levels. #UDLsnowdays

FLIPPED CLASSROOMS

The flipped classroom model sounds promising, yet as of 2013, there was no scientific research base to indicate the effectiveness of flipped class-rooms (Goodwin & Miller, 2013). As with other instructional practices, however, when that research base is built, we predict the experiences of students in flipped classrooms will be wildly different depending on how universally designed their online experiences are.

Many faculty record their lectures (either in class or in their offices) and make the videos available from their LMS or YouTube account and ask their students to view the videos before coming to class so that the students will be ready to discuss and apply the new content. Other instructors use tools such as Khan Academy videos and encourage students to attend virtual classes online. Educause (2012), the leading organization for the use of technology in higher education, states that the flipped classroom is significant because it allows students to customize the pace of instruction. For example, in a traditional lecture, students cannot stop the instructor from talking as they take notes and reflect on what is being said. When watching video, however, the students control the pace of the lecture.

Since students' first exposure to materials is done online, class time gives them an opportunity to use higher cognitive skills as they work to apply knowledge in collaboration with peers and the instructor. Flipped classrooms, in theory, improve the outcomes of student learning because students have more support and mastery-oriented feedback when they struggle with more difficult tasks. As with any instruction, however, if these flipped classrooms are not universally designed, improved student outcomes are unlikely.

If you wish to create lecture videos and flip your classroom, keep the videos short, 5 to 10 minutes, and integrate other active assignments like a discussion board or formative assessment into the process to make them easier for the students to watch and learn from—just like you would do in a face-to-face classroom. One way to do this is to chunk the content into small video segments that you surround with activities (Bodie, et al., 2006). A short video is a great way to introduce a topic and demonstrate a process, but as we stated previously, the optimum length for an online video is 6 minutes. (What would Ben say about that?)

Also, don't expect that students will take to this new form of instruction without explicit scaffolding. Students who have never participated in a flipped classroom may feel overwhelmed by the need to highlight patterns, critical features, big ideas, and relationships independently, since this is often done with support from the instructor in the classroom. After receiving direct instruction, students will need opportunities to collaborate with peers to learn strategies for processing information, taking notes, and managing all the resources posted online before they can fully access the flipped classroom from home.

> **@PoorRichardUDL** · To rephrase Ben: A video short is a video learned. #flippeddoesn'tequallearned

THE VIRTUAL SNOW DAY

The virtual snow day is a close cousin of the flipped classroom. A virtual snow day would be used any time that students are unable to come to school, typically when the school is closed in a weather emergency. Some virtual snow days run as a collection of flipped classrooms, where students watch a series of videos that replace classroom lecture and complete assignments in an LMS. Other virtual snow days might consist of live video conferences that connect the teacher from his or her home office to each student's home through technology like the Big Blue Button, Blackboard Collaborate, Cisco WebEx, Citrix GoToMeeting, or other application that mimics the use of a virtual meeting like many businesses now use. But, if this has any chance of being successful we will all need to ensure a few things:

1. All students will need compatible technology at home and in school.

2. All students will need to have a high-speed Internet connection to access an LMS or other applications.

3. The school system or college must have a system in place to deliver online content (an LMS like Canvas).

4. The faculty and students must be trained on how to use the system.

5. The educational content of the snow day must be available in the system.

6. Students and teachers will have to use the LMS (or other application) on some sort of a regular basis as part of the everyday process of education (in class) so that they will be ready to use it online when they are not in class (when it snows).

Let's look at each item individually.

All students will need compatible technology at home. A school system needs to ensure that all students have devices, but this cost is significant. In response, some school departments and colleges are working toward one-to-one programs that include the technology in the cost of tuition or where it is supplied as part of the capital technology plan (Thompson, 2014).

This setup is ideal, but many institutions and school districts will not be able to afford this technology. The digital divide is already wide, and getting wider, and access to the technology is just the proverbial "tip of the iceberg." If the school district cannot ensure that all students have access to technology, and have the skills to use it appropriately, virtual snow days will not be successful.

All students will need to have a high-speed Internet connection to access an LMS or other applications. Once school districts are confident that students have available technology, they must then be sure that all students can connect to a network that is sufficient to deliver the content and videos needed to replicate an in-class experience. In some areas of the country, this might not even be possible yet due to the costs of the system installation. Universal broadband may eventually make this possible, but we are still years away from such an implementation. Students without a broadband connection are definitely on "the other side of the tracks" in the digital divide.

According to Netflix you need the following download speeds to stream video:

> ▶ **Internet Connection Speed Recommendations**
>
> Below are the Internet download speed recommendations per stream for playing movies and TV shows through Netflix.
>
> - 0.5 megabits per second (Mbps)—Required broadband connection speed
>
> - 1.5Mbps—Recommended broadband connection speed
>
> - 3.0Mbps—Recommended for standard-definition (SD) quality
>
> - 5.0Mbps—Recommended for high-definition (HD) quality
>
> - 25Mbps—Recommended for Ultra HD quality

According to the National Broadband Map (NTIA, 2015), 95.4% of the population has that ability, but there are still large areas where students would not be able to access course content.

SPEED	NATIONWIDE
Download >3Mbps; Upload > 768Kbps	94.8%
Download > 3Mbps	95.4%
Download > 6Mbps	94.2%
Download > 10Mbps	92.9%
Download > 25Mbps	85.3%
Download > 50Mbps	83.2%
Download > 100Mbps	64.8%
Download > 1Gbps	7.9%

The school system or college must have a system in place to deliver online content. Trying to create a successful virtual snow day in a school district or college without the help of an LMS would be challenging. Instructor expectations would vary greatly and students would have to navigate significant differences without any support. Having an LMS allows all staff to receive the same professional development when learning how to deliver online instruction, so students can transfer this knowledge as they "switch" classes and advance to different grades. Having the same system also allows students to become more proficient as time goes on and allows districts to involve parents and the community in setting expectations for the LMS's use.

Faculty and students must be trained on how to use the system. It is often stated that today's students are very technologically capable. If you watch many of these students texting or playing an online video game, you will certainly see a level of that technological proficiency. However, that proficiency does not always transfer to an educational task without the same training and practice that was invested in the texting or playing. This training will take time. Instructors will have to be trained first, and they will need to be given time to implement that training into their courses in the LMS. They will also have to train their students on how to use the LMS. Professional development plans will have to be created and implemented, and these plans may require additional negotiations regarding union contracts.

The educational content of the snow day must be available in the system. In order for a virtual snow day to work, the process has to work for both the teacher and the student. In order for that to happen, the content has to be in the LMS ready to go. Teachers may not be able to put a whole day's worth of student work into the LMS after the cancellation of the school day at 6:00 a.m. They might be able to put *something* into the system, but it wouldn't be an exemplary lesson equal to a traditional face-to-face class.

Students and teachers will have to use the LMS on some sort of a regular basis as part of the everyday process of education (in class) so that they will be ready to use it online when they are not in class (when it snows). The LMS has to be an integrated component of the course. It will need to be populated with daily content; it will have to be used as part of the daily set of activities in the class. Obviously, the classrooms have to support the technology, too. Every class will need a computer and projector. The Wi-Fi in the buildings will have to be robust and the network that supports it will have to be fail-safe. It all has to work 99.9% of the time. Any weak link has to be eliminated, or as Murphy states, "it will fail when you need it the most"— the first snow day of the year. If that happens, you will lose precious momentum and faculty support.

This will require that we set up our classrooms for "anywhere, anytime learning." Remember, we cannot change the incoming capabilities of our students. We can only react to those capabilities and build from there. That means that we have to find ways to maximize structured time on task, maximize student engagement outside of class time, and try to make the technology as transparent as possible—without being physically there.

To do that we have to make sure that we understand good course design before we load our courses onto the cloud. Good courses take time to create and deliver.

As we have mentioned, the best way to manage all of this is with a LMS. An LMS is most effective if it is leveraged heavily by connecting the in-class experience to the out-of-class experience. At its best, an LMS gives the instructor a "place" to develop, deliver, and manage the content and experience of a course or classroom.

Using the LMS in class will reinforce its use out of class. If "in-class" and "out-of-class" merge into a new, different dynamic, you can ask that students acquire content and complete class work outside of class, which will allow for more UDL choice.

An LMS can greatly enhance an instructor's ability to flip the classroom, too, so teachers won't have to use class time to lecture and they can spend more time on student understanding, interaction, implementation, assessment, and reflection. This will allow students to get actively involved in class preparation and significant projects connected to the real world. Finally, if instructors can get the experience necessary to work in both environments, they will know exactly what changes are required to make a fully online course interesting and effective.

@PoorRichardUDL · It will take more than snow to derail a well-constructed UDL lesson. #Letitsnow

REFLECTION

- If you are currently teaching a hybrid-online course, or exploring the option of teaching one, like a virtual snow day, it's important to provide students with options for comprehending the online content while students are in a face-to-face class, so they are prepared when they have to access it independently. What are some concrete ways you can use an LMS or Google Drive in your classroom to support students?

OPTIONS FOR EXPRESSION

- If your institution doesn't have an LMS, complete some research on various platforms to see what they are capable of. Some popular LMSes are Blackboard, Canvas, and itslearning. Prepare a presentation or paper to share with stakeholders at your institution about the benefits of adopting an LMS.

- If your institution has an LMS, design a snow day or a flipped classroom using as many of the tools as possible, and then deliver it to students, face-to-face, while in a computer lab or while working on 1:1 devices. Make observations and take notes on the areas that are confusing and collaborate with students to make tutorials they can access from home.

- If adopting an LMS is not an option at your institution, explore Google Classroom or the free version of Canvas that offers a free course to all teachers. Create a test module and share with your students so together you can develop a list of guidelines that will allow them to be successful in that environment.

8

Giving Our Students the Final Word

IN THIS CHAPTER, we will return to the profiles presented in Chapter 1 and provide a different picture of the students, assuming instructors implemented a universally designed course. Readers can explore how changes to course development, delivery, instructor presence, and time management can transform a course and student outcomes and can move online education away from the problems that may arise.

Let's see how a universally designed course, aligned to the UDL Guidelines, would impact our students. Even though they are vastly different, you will see that using the strategies in this text will allow you to build a course that will meet all their needs. From designing the syllabus, to building instructor presence, to scaffolding important executive functions and time management skills, to delivering and evaluating your course—UDL is the framework that will allow you to meet the needs of every student in the cloud.

COCO

Coco was worried about getting ready for the first virtual snow day, but when her teacher integrated the LMS into the classroom on a daily basis, she learned how to use the system. She completed a time-management assignment with the help of her parents, and realized that she could share the computer with her older siblings and they could all do the work that they were required to do during the day.

UDL GUIDELINE	IMPLICATIONS FOR COCO
Provide options for self-regulation • Promote expectations and beliefs that optimize motivation • Facilitate personal coping skills and strategies • Develop self-assessment and reflection	Before the first snow day, Coco's teacher used the LMS in the classroom on the Smart Board, so all students were familiar with how the virtual snow days would be organized. The teacher also helped her students develop strategies for using the LMS by taking her class to the computer lab on a regular basis and showing them how to navigate the system. She integrated the LMS into daily class activities. The teacher also posted rubrics on the LMS so the students understood exactly how their work would be graded, even though the teacher wasn't present to answer questions.
Provide options for sustaining effort and persistence • Heighten salience of goals and objectives • Vary demands and resources to optimize challenge • Foster collaboration and community • Increase mastery-oriented feedback	Every time Coco logged on to the LMS, the first screen included an announcement with the goals and objectives for the day, organized as a to-do list. The teacher also encouraged students to post drafts of their work in the discussion board so she, and the other students in the class, could make suggestions for revision and provide feedback.

UDL GUIDELINE	IMPLICATIONS FOR COCO
Provide options for recruiting interest • Optimize individual choice and autonomy • Optimize relevance, value, and authenticity • Minimize threats and distractions	Coco always had choices on virtual snow days, which she began to enjoy. She liked that she could read an article, work on math problems in her book, and watch a video while cozying up in her pajamas with hot cocoa.
Provide options for executive functions • Guide appropriate goal-setting • Support planning and strategy development • Enhance capacity for monitoring progress	Before the first snow day, Coco's teacher assigned each student the task of creating the schedule they would follow on the first day, and included meals, computer time, and time to build a snowman. Coco worked with her family to set up a schedule for her and her brothers to use the computer. Coco also always printed the to-do list and crossed off tasks as she completed them.
Provide options for expression and communication • Use multiple media for communication • Use multiple tools for construction and composition • Build fluencies with graduated levels for support for practice and performance	Coco's teacher allowed students to print out reading assignments and articles from the LMS, and she always had an assignment from their textbook, so if students couldn't be on the computer all day, they could still complete the work. The teacher also made herself available in a chat feature in the LMS during regular school hours so students could contact her if they had a question.

UDL GUIDELINE	IMPLICATIONS FOR COCO
Provide options for physical action • Vary the methods for response and navigation • Optimize access to tools and assistive technologies	When posting a prompt that required a response, Coco's teacher encouraged students to type written reflections, record audio on smart devices, or write assignments on paper and turn them in the following day. This minimized Coco's anxiety since she didn't have to worry about whether she had the time to type all her responses on the shared computer.
Provide options for comprehension • Activate or supply background knowledge • Highlight patterns, critical features, big ideas, and relationships • Guide information processing, visualization, and manipulation • Maximize transfer and generalization	The teacher activated the students' background knowledge before the first snow day occurred. This repeated practice allowed them to transfer these skills when they worked independently during the storm.
Provide options for language, mathematical expressions, and symbols • Clarify vocabulary and symbols • Clarify syntax and structure • Support decoding of text, mathematical notation, and symbols	The structure of the course was clarified for the class in the computer lab before Coco had to use the computer. The teacher also had a parent information night where she introduced the LMS to parents so they could support their children if they had issues navigating the site.

UDL GUIDELINE	IMPLICATIONS FOR COCO
Provide options for perception • Offer ways of customizing the display of information • Offer alternatives for auditory information • Offer alternatives for visual information	For each snow day module, the teacher posted a list of options on the to-do list so students could read an article or book, watch a video, or view a short PowerPoint lecture. Coco often chose to read, so she could complete her work while her brothers logged on to their LMS.

KRITI

Kriti had a positive experience in her virtual course because of her connection with the instructor. The instructor built a strong presence and reached out to her young students using the technology they were most comfortable with, in order to build important skills they could then transfer to experiment with different 21st-century technologies.

UDL GUIDELINE	IMPLICATIONS FOR KRITI
Provide options for self-regulation • Promote expectations and beliefs that optimize motivation • Facilitate personal coping skills and strategies • Develop self-assessment and reflection	One of the first assignments in the course required Kriti to watch videos and review short research articles about how to succeed in an online course and then respond to a prompt on the discussion board about why online learning was important. She shared this assignment with her parents and this began to break down the cultural barriers she faced.
Provide options for sustaining effort and persistence • Heighten salience of goals and objectives • Vary demands and resources to optimize challenge • Foster collaboration and community • Increase mastery-oriented feedback	Kriti's instructor created a very complete and detailed syllabus and then used an adaptive release feature so Kriti was only presented with one assignment at a time, which allowed her to focus on the goals for the day. Also, the teacher encouraged students to share their contact info with each other in an early discussion post, including their Facebook, Twitter, Pinterest, and Instagram handles and their personal emails so they could collaborate when necessary.
Provide options for recruiting interest • Optimize individual choice and autonomy • Optimize relevance, value, and authenticity • Minimize threats and distractions	At the end of each unit, the instructor shared the objectives for the next unit and asked her students to make suggestions about topics they'd like to study or activities they would like to complete as they worked toward the unit objectives.

UDL GUIDELINE	IMPLICATIONS FOR KRITI
Provide options for executive functions • Guide appropriate goal-setting • Support planning and strategy development • Enhance capacity for monitoring progress	Kriti's instructor was always available via email and phone and had "office hours" every afternoon where she would text students who were struggling to help get them back on track. Kriti felt very comfortable with this form of communication and this encouraged her to stay on track. Also, one assignment a week required students to create a plan for completing their work. This weekly exercise helped Kriti to build important self-regulation strategies.
Provide options for expression and communication • Use multiple media for communication • Use multiple tools for construction and composition • Build fluencies with graduated levels for support for practice and performance	Kriti found the support she needed from her instructor by taking advantage of the support person in her school library and the virtual office hours for her first few weeks of class. This support was crucial for her to understand what was expected of her. Also, the frequent check-ins from her instructor allowed her to stay focused.

UDL GUIDELINE	IMPLICATIONS FOR KRITI
Provide options for physical action • Vary the methods for response and navigation • Optimize access to tools and assistive technologies	Kriti loved that she was encouraged to access instruction on her smartphone and iPad. She came to rely on her iPad at all hours of the day and night and eventually knew the technology so well that she was able to create a video essay as her final project.
Provide options for comprehension • Activate or supply background knowledge • Highlight patterns, critical features, big ideas, and relationships • Guide information processing, visualization, and manipulation • Maximize transfer and generalization	Each week, the instructor highlighted the big ideas for the week and activated an adaptive release feature so additional videos, texts, and assignments weren't a distraction to her students if they weren't ready for the additional content.
Provide options for language, mathematical expressions, and symbols • Clarify vocabulary and symbols • Clarify syntax and structure • Support decoding of text, mathematical notation, and symbols	Since Kriti has a difficult time paying attention, she benefited from the multiple representations of all course content. She transitioned from reading, to watching videos, to viewing a Power-Point as she worked to comprehend difficult topics across the content area.

UDL GUIDELINE	IMPLICATIONS FOR KRITI
Provide options for perception • Offer ways of customizing the display of information • Offer alternatives for auditory information • Offer alternatives for visual information	The instructor posted a short, funny video each day to introduce the day's lessons. Kriti especially loved the videos in the course because she was able to start and stop them as she needed and rewind and replay them whenever she was unclear.

JUNE

It took June a little longer to do things in the beginning of the course, but because of the scaffolding and built-in supports, she quickly acclimated to the tasks.

UDL GUIDELINE	IMPLICATIONS FOR JUNE
Provide options for self-regulation • Promote expectations and beliefs that optimize motivation • Facilitate personal coping skills and strategies • Develop self-assessment and reflection	June loved having an instructor who was available to answer questions and virtually hold her hand through early problems of connectivity. Also, she loved that the instructor had in-person office hours for students who were geographically close. The checklists, rubrics, and clear instructions for all assignments allowed June to flourish.
Provide options for sustaining effort and persistence • Heighten salience of goals and objectives • Vary demands and resources to optimize challenge • Foster collaboration and community • Increase mastery-oriented feedback	She found a great sense of community in the group work and discussion board. She grew comfortable asking her classmates for assistance early on in the program and ended up being able to share her years of experience on teaching and writing when she got over her initial fears of the computer tasks.
Provide options for recruiting interest • Optimize individual choice and autonomy • Optimize relevance, value, and authenticity • Minimize threats and distractions	Being able to print out assigned reading provided June with a traditional connection that she was comfortable with and gave her a learning option that was not "online." Also, since all course assignments were things she could use in designing her own course, she felt as though every minute she spent on homework was a good use of time.

UDL GUIDELINE	IMPLICATIONS FOR JUNE
Provide options for executive functions • Guide appropriate goal-setting • Support planning and strategy development • Enhance capacity for monitoring progress	June loved that the goals for each week were clear, and that the instructor encouraged all students to add assignments to their personal calendars so they could monitor their progress.
Provide options for expression and communication • Use multiple media for communication • Use multiple tools for construction and composition • Build fluencies with graduated levels for support for practice and performance	June needed lots of tutorials and support so she loved all the background text and research articles the instructor posted as options. Also, she always reviewed the exemplars for each assignment so she knew exactly what was expected of her.
Provide options for physical action • Vary the methods for response and navigation • Optimize access to tools and assistive technologies	June prefers to navigate through information and activities in more traditional ways, but that is because that is where her comfort level is. In an early assignment, the instructor required the students to explore at least one new assistive technology and reflect on the experience. The teacher provided a list of tools to explore and June was so engaged she wanted to try them all.

UDL GUIDELINE	IMPLICATIONS FOR JUNE
Provide options for comprehension • Activate or supply background knowledge • Highlight patterns, critical features, big ideas, and relationships • Guide information processing, visualization, and manipulation • Maximize transfer and generalization	Having a syllabus that is a true representation of the course requirements was a treasure for June, because she could print it, follow along, check off what she had completed, and look ahead. This highlighted the big ideas of the course and allowed her to build technological competencies as the weeks went on.
Provide options for language, mathematical expressions, and symbols • Clarify vocabulary and symbols • Clarify syntax and structure • Support decoding of text, mathematical notation, and symbols	June has very strong comprehension skills so she is adept at decoding text. Once her instructor learned this, he encouraged June to create a presentation about how to decode difficult text as an assignment, so she could showcase her skills to her peers.
Provide options for perception • Offer ways of customizing the display of information • Offer alternatives for auditory information • Offer alternatives for visual information	June succeeded in the course because she always had the option to read text to learn new content. This gave her the confidence to brave new computer technologies.

JOSÉ

José will probably succeed in any educational environment. He follows directions and participates freely in his classes. He loves online courses for the flexibility it gives him to choose when and where he can do his course work. But, he especially likes the options presented to him in an UDL-compliant course that allow him to change the way he does his course work.

UDL GUIDELINE	IMPLICATIONS FOR JOSÉ
Provide options for self-regulation • Promote expectations and beliefs that optimize motivation • Facilitate personal coping skills and strategies • Develop self-assessment and reflection	José is an autonomous, self-regulated learner already, so it was important for his instructor to provide options that include significant levels of challenge to keep José engaged.
Provide options for sustaining effort and persistence • Heighten salience of goals and objectives • Vary demands and resources to optimize challenge • Foster collaboration and community • Increase mastery-oriented feedback	In addition to providing various levels of challenge, the instructor encouraged him to connect with others in collaborative assignments and on the discussion board. José liked building relationships with his peers and especially liked how they noted weaknesses in his work, because this pushed him to work harder so his future employers wouldn't find the same weaknesses. Lastly, the instructor built a strong presence and José respected the instructor's expertise. Because of this, José was receptive to the instructor's feedback, which took José's work to the next level.

UDL GUIDELINE	IMPLICATIONS FOR JOSÉ
Provide options for recruiting interest • Optimize individual choice and autonomy • Optimize relevance, value, and authenticity • Minimize threats and distractions	José's instructor always posted a mini-lecture where he explained the relevance of the week's assignments. These videos spoke to the importance of learning the new concepts and reinforced José's need to improve.
Provide options for executive functions • Guide appropriate goal-setting • Support planning and strategy development • Enhance capacity for monitoring progress	José doesn't like to waste his time, so he loved the clarity of the syllabus because it was clear what he needed to do to succeed. José handed in all his assignments early, and his work was proficient, and so he already had the skills to monitor his own progress. As a result, his instructor focused on developing a connection with José to keep him engaged.
Provide options for expression and communication • Use multiple media for communication • Use multiple tools for construction and composition • Build fluencies with graduated levels for support for practice and performance	José loved the assignments, which required real-world application, where he was encouraged to choose from a list of technologies or others of interest to him.

UDL GUIDELINE	IMPLICATIONS FOR JOSÉ
Provide options for physical action • Vary the methods for response and navigation • Optimize access to tools and assistive technologies	The instructor did not let José get too comfortable with just one way of responding on the discussion board. Instead, he encouraged all students to respond differently each week if possible. This kept José interested and pushed him to try new things.
Provide options for comprehension • Activate or supply background knowledge • Highlight patterns, critical features, big ideas, and relationships • Guide information processing, visualization, and manipulation • Maximize transfer and generalization	José wants real-world application. Because he has strong background knowledge, he feels it's a waste of time to review concepts he already knows. That's why he liked the options of watching videos or reading text because he could skim through the parts he already knew.
Provide options for language, mathematical expressions, and symbols • Clarify vocabulary and symbols • Clarify syntax and structure • Support decoding of text, mathematical notation, and symbols	José can handle new vocabulary, syntax, and symbols most of the time, so he appreciated that he could choose to view background information or review important vocabulary ahead of time, but it was not required. He was glad he didn't have to complete busy work for no reason.

UDL GUIDELINE	IMPLICATIONS FOR JOSÉ
Provide options for perception • Offer ways of customizing the display of information • Offer alternatives for auditory information • Offer alternatives for visual information	Even though José didn't require any of these options, he really liked having them, because they increased his engagement in the course.

RAY

Ray is a hard worker trying to balance work, family, and school. To be successful, he needs a course with a clear syllabus that has all the expectations of the course mapped out from the beginning. Having an instructor who is available and responsive is also crucial because Ray only has a certain amount of time to complete assignments and so he can't wait around for answers or feedback.

UDL GUIDELINE	IMPLICATIONS FOR RAY
Provide options for self-regulation • Promote expectations and beliefs that optimize motivation • Facilitate personal coping skills and strategies • Develop self-assessment and reflection	Setting assignment expectations in the syllabus kept Ray on track. In addition, setting deadlines and estimates of the time necessary to do the assignments helped Ray manage his time accurately and do the work. Providing time for reflection within the discussion board and regular formative assessment opportunities helped Ray realize what he was learning and what it would mean to his future, which allowed him to persist when he was exhausted.
Provide options for sustaining effort and persistence • Heighten salience of goals and objectives • Vary demands and resources to optimize challenge • Foster collaboration and community • Increase mastery-oriented feedback	Repeating the goals of the course each week and then seeing the application of the goals helped Ray sustain his efforts. Giving Ray multiple modes of mastery-oriented feedback helped him to realize what he was doing well and what he needed to spend more time on in order to improve. Also, because the instructor spent time building his presence, Ray felt comfortable reaching out to him for feedback throughout the course.

UDL GUIDELINE	IMPLICATIONS FOR RAY
Provide options for recruiting interest • Optimize individual choice and autonomy • Optimize relevance, value, and authenticity • Minimize threats and distractions	The instructor always provided students with a rationale for each assignment and why it was important. This helped Ray to connect all assignments to his future career. Ray received positive, mastery-oriented feedback, which also supported his progress and minimized his feelings of inadequacy because of his weak educational background.
Provide options for executive functions • Guide appropriate goal-setting • Support planning and strategy development • Enhance capacity for monitoring progress	Ray appreciated that the course syllabus outlined the time students should set aside for each task. This was important for Ray so he could plan his weeks around work and family. Also, the LMS had an early warning feature set up, so if Ray ever posted an assignment late, the system notified him and the instructor right away. At this point, the instructor always reached out and helped Ray design a plan to get back on track.
Provide options for expression and communication • Use multiple media for communication • Use multiple tools for construction and composition • Build fluencies with graduated levels for support for practice and performance	Again, since Ray was always starting schoolwork after work, it was good that there were always different options and tools he could use to keep him interested. Monotonous assignments would be far too difficult for him to complete, since he was already tired.

UDL GUIDELINE	IMPLICATIONS FOR RAY
Provide options for physical action • Vary the methods for response and navigation • Optimize access to tools and assistive technologies	Since Ray had to begin all class activities after a full day of work, activities that required some sort of physical action reinvigorated him. For example, at times he was encouraged to shoot videos, which got him out of his chair to record the action around him. Providing a variety of tools peaked his interest (since he is good with computers) and showed him how he can better use technology to move forward.
Provide options for comprehension • Activate or supply background knowledge • Highlight patterns, critical features, big ideas, and relationships • Guide information processing, visualization, and manipulation • Maximize transfer and generalization	Ray needed a UDL-compliant course to help him manage the time commitment required in his courses. Since everything was laid out for him in the syllabus and in the weekly tasks, he was able to make the time to do the work, and do it successfully. More importantly, the course content was represented in several different ways that allowed him to choose the best mode for the time that he had available and the complexity the mode required.
Provide options for language, mathematical expressions, and symbols • Clarify vocabulary and symbols • Clarify syntax and structure • Support decoding of text, mathematical notation, and symbols	Ray is not a glossary reader. He wants to know what everything means as he approaches it. The instructor supported him by never using jargon or unfamiliar terminology without carefully explaining it.

UDL GUIDELINE	IMPLICATIONS FOR RAY
Provide options for perception • Offer ways of customizing the display of information • Offer alternatives for auditory information • Offer alternatives for visual information	Ray appreciated having the option to watch a video to learn new content after a particularly long day at work. He also loved the audio option for class reading, so he could listen to it on his iPhone while preparing dinner for his family.

Like these five hypothetical students, all of our students are different—whether they are in traditional face-to-face classes, hybrid-online courses, a virtual snow day, or an online course. Our job as educators is to help each of them to learn. Ben Franklin, once again, may have said it best: "Without continual growth and progress, such words as *improvement, achievement,* and *success* have no meaning." Universal Design for Learning is a next step in that process.

@PoorRichardUDL • Growth mindset + UDL = perfect combination for success. #makeUDLhappen

REFLECTION

- Think about a student you have taught who struggled in your hybrid-online or online course, or dropped the course before completion. Review the UDL Guidelines and reflect on the content of this book. How would your student's outcomes have been different if your online or hybrid-online course was universally designed?

- What is one strategy that you will begin to implement immediately to ensure that all students can come one step closer to succeeding in your online environment?

OPTIONS FOR EXPRESSION

- Build an online module or course to support other instructors in universally designing their own online courses or virtual snow day. Use the research and best practices outlined in this text to design a syllabus and develop the course. Application of this knowledge will help you to reflect on the material and will give you an opportunity to implement some suggestions as you improve online courses or virtual snow days at your institution.

- Review each chapter of this book and create a short summary of each chapter. Keep this resource handy as you develop your own online module or course.

- Review the UDL Guidelines and then examine an online course or virtual snow day you currently teach. Assign your course a grade (A–F) for each Guideline so you know which aspects of the course you will need to focus on as you make revisions.

- Write a blog post or book review about *UDL in the Cloud* and share it with us. You can find us on Facebook, Twitter, LinkedIn, and at www.katienovakudl.com. We welcome your feedback any time.

Appendix

Pedagogy and Andragogy: Why K–12 and Adult Learners Aren't So Different

IN THIS APPENDIX we explore the learning theory of andragogy and how it purportedly differs from pedagogy. This research will be juxtaposed with UDL theory to elucidate the need for the UDL Framework when teaching learners of all ages.

As we mentioned in the text, there are colliding theories that argue that children and adults learn differently: pedagogy and andragogy. Although andragogy is strongly recognized in the literature, it has come under fire over the last decade, and for good reason. First, critics argue that andragogy lacks the fundamental characteristics of a science because it cannot be measured (Taylor & Kroth, 2009).

Second, others argue whether or not andragogy is a theory at all. Critics argue that andragogy is just a model of assumptions about how adults learn best. Knowles (1989), the modern father of andragogy, has even said that andragogy is less of a theory and more a "model of assumptions about learning or a conceptual framework that serves as a basis for an emergent theory" (p. 112).

Third, because both the design and delivery of curriculum are collaboratively determined by the instructor and learner, there are no set practices to measure.

Fourth, "Knowlesian andragogical effectiveness" is largely determined by learner achievement, which is often measured by tests and grades; but for Knowles, "tests and grades are anathema to the very idea of andragogy" (Taylor & Kroth, 2009, p. 8).

Lastly, and we believe most importantly, many of the characteristics of adult learners may also be present in children, therefore blurring the lines between andragogy and pedagogy.

First, these two colliding theories suggest that adults learn differently from children, or K–12 students, but evidence shows that the reality is much broader—everyone learns differently. Age is certainly a factor in the process, but there are many other variables in the learning process that can be more important than an individual's age. The brain has a unique blend of network interactions that defines individual learning from birth (Meyer, Rose, & Gordon, 2014). If you consider that all learners are different, age may exacerbate that variability, but regardless of age, learners are different.

So, what does the theory of andragogy argue is unique about adult learners and how is the education of adults different from educating K–12 learners? Adult learners come to the classroom with much of the same variation as K–12 learners, with both learning and school readiness (Zafft, Kallenbach, & Spohn, 2006). Their life experiences, however, amplify both potential educational issues and their potential solutions (Knowles, 1984; Knowles, Holton, & Swanson, 1998).

Because these life experiences have the potential to increase variation, much research has focused on how adults learn best. Knowles (1984) made the following assumptions about pedagogy and then offered andragogy as a stronger model. We want to note that these assumptions made by Knowles seem rather dated when examining research on UDL and best practices, but it's important to provide the background to understand how the theory of andragogy came to be.

Three Assumptions about Pedagogy (Knowles, Holton, & Swanson, 1998)

1. Students only need to learn what the teacher teaches them to answer questions during an exam.

2. Prior experience is not necessary in pedagogy.

3. Students have no knowledge in a particular area and therefore they have to depend solely on the teacher to learn the basics.

If these assumptions were indeed true, authentic learning would be unlikely for K–12 students or adult learners. However, Knowles's focus was on adults, and he proposed six assumptions regarding the characteristics of adult learners, arguing that pedagogy, given his three assumptions, would not meet their needs.

Six Characteristics of Adult Learners (Knowles, 1980)

1. Adults tend to see themselves as more responsible, self-directed, and independent.

2. They have a larger, more diverse stock of knowledge and experience.

3. Their readiness to learn is based on developmental and real-life responsibilities.

4. Their orientation to learning is most often problem-centered and relevant to their current life situation.

5. They have a stronger need to know the reasons for learning something.

6. They tend to be more internally motivated.

Supporters of andragogy outline multiple benefits of focusing on these six assumptions when teaching adults. For example, when using the andragogical principles, the instructor can personalize instruction to maximize student interest and engagement. In addition, andragogy purportedly increases collaboration and communication between adult students and instructors as they work together to design learning experiences that are relevant, meaningful, and authentic (Chan, 2010).

Knowles (1984) argues that because adults bring a diverse combination of knowledge, experience, and independence to the classroom, adult educators should work to ensure that adult learners participate as much as possible in the content, delivery, and evaluation of curricula within a climate of mutual respect.

In addition, in a review of research on andragogy, McGrath (2009) noted that andragogy allows adults to make connections between teaching, learning, and their own lives while pedagogy is focused on the students learning "word for word so that they can receive positive feedback from their lecturers" (p. 102).

To provide a graphic representation of the purported differences between pedagogy and andragogy, examine the following table (constructed from Knowles, 1984).

COMPARISON OF PEDAGOGY AND ANDRAGOGY

	PEDAGOGICAL	ANDRAGOGICAL
Concept of Learner	Responsibility of instructor	Student-directed with resistance to being "told"
Role of Learner's Experience	Minimized—previous grade	Maximized (part of self-identity) and expected to be part of the process of learning new information
Readiness to Learn	Largely a function of age	Ready to learn based upon need to improve
Orientation to Learning	Subject-oriented and sequenced to logic of subject matter	Life-, task-, or problem-oriented; sequenced to the logic of the life situation: "need to know," "relevant"
Motivation to Learn	External	Internal
Format	Content plan	Process design
Instructor Role	"Sage-on-the-stage"	"Facilitator"
Learning Expectation	Predetermined by standard	Potentially open-ended

We'd like to examine these differences using the UDL lens and current research on instructional best practices, since the theory of pedagogy has evolved to meet the needs of all learners, and the assumptions of Knowles in the previous table do not account for that. When supplementing the proposed model of pedagogy with recent research that supports UDL as a best practice, pedagogy and andragogy are more tightly related than Knowles's assumptions suggest.

First, Knowles (1984) suggests that in the theory of pedagogy, the **concept of the learner** is the responsibility of the instructor. According to self-determination theory, however, learners experience autonomy, and are internally motivated to learn when they can realize their personal goals, values, and interests and see how they connect to learning. Instructors can maximize this motivation by providing choice and acknowledging the learner's perspective and feelings, but instructors do not create this autonomy (Assor, Kaplan, & Roth, 2002; Sanacore, 2008), and this autonomy is not a function of age.

Knowles also argues that the **role of a learner's experience** is minimal, and is dependent on the previous grade of students—not life experiences. This is simply not true. For example, a recent study examined the role of poverty on children's experiences in school. The authors found that growing up in poverty significantly impacted a children's development—especially their ability to reach their full learning potential (Hilferty, Redmond, & Katz, 2010). If the role of a child's experience was only a function of previous grade, confounding variables, such as poverty, would not impact students' ability to access teaching and learning.

Similarly, Knowles (1984) notes that an adult's **readiness to learn** is based on developmental and real-life responsibilities, and argues that children's readiness is largely a function of age. Contemporary researchers argue that readiness to learn is a shared responsibility that is both interactive and relational—and begins immediately after birth. Therefore, learners of all ages are always "ready to learn" (Hilferty, Redmond, & Katz, 2010). The distinction, they argue, is that sometimes students may not be ready for school because they lack the ability to interact effectively in the classroom, listen with attentiveness, and follow simple directions. This does not, however, affect a child's readiness to learn.

Conversely, an adult's readiness to learn can be affected by variables such as stereotype threat. Stereotype threat is the phenomenon in which individuals underperform on academic tasks when subtly or subconsciously reminded of their group membership before beginning the task (Heerboth & Mason, 2012). Research has shown that stereotype threat can harm the academic performance of any individual for whom the situation invokes a stereotype-based expectation of poor performance. Therefore, adult learners bring many of the same educational issues into the classroom as younger learners do.

When considering **orientation to learning**, Knowles (1984) notes that adult education must be relevant and "life"-oriented instead of subject-oriented. With the adoption of college and career readiness standards nationally, it is clear that the instruction of students K–12 incorporates 21st-century skills that are important for "life." In fact, the Common Core State Standards, which have been adopted by 42 states, were developed by "carefully studying the research and literature available on what students need to know and be able to do to be successful in college, career, and life."

When examining **motivation to learn**, pedagogy is described as focusing on the external motivation of students; however, it is well documented that relationships between self-regulated learning strategies and motivational beliefs increase student engagement, student attitude, student motivation, and student performance in different grades and with different lessons (Ocak & Yamac, 2013).

Another difference outlined in Knowles's comparison of pedagogy and andragogy is the **format of lessons**. Pedagogy is described as being organized in a content plan, whereas andragogy is a process design. The UDL Framework, however, cautions against the rigid format of lessons, in content or process. When considering curricular design, Rose & Meyer (2006) ask, "What if we recognized that our inflexible curricula and learning environments are 'disabled' rather than pinning that label on learners who face unnecessary barriers?" (p. vii). Any format, therefore, that does not consider the significant variation of learners is flawed from the beginning.

The **role of the instructor** is also different, with the two ends of the spectrum being "sage-on-the-stage" in the pedagogical model to "facilitator," or "guide-on-the-side" in the andragogical model. In order for all learners to maximize their potential and learn 21st-century skills, all teachers need to be experimental and error-welcoming to engage students. In short, all educators need to facilitate learning experiences, or coach learners in the learning environment. The value of such activity is strongly supported by findings from neuroscience about student engagement. All educators must shift to value creation rather than knowledge transmission (McWilliam, 2008).

Lastly, Knowles (1984) notes a difference in the **learning expectations** in the two theories, since pedagogy focuses on standards-based learning. Recent research supports the importance of standards-based instruction for both K–12 students and adults by noting the value of identifying a set of learning objectives or "standards" for each course and then grading students on the mastery of those standards (Beatty, 2013). If a course does not have clear learning objectives or standards, there is the possibility of instructors having different standards for different learners, which can influence the quality of education and exacerbate stereotype threat.

When an instructor has standards, or intended outcomes, for instruction, they are less likely to look toward individual students to provide entry points, learning tasks, and outcomes, which allows those students to access the same rigorous curriculum (Hall, Strangman, & Meyer, 2003). When instructors leave course outcomes open-ended, there may be a tendency to create different levels of expectations for learners so they can be successful (Waldron & McLeskey, 2001). Different levels of expectations create an unequal education.

The previous discussion attempts to minimize the gap between andragogy and pedagogy, since both theories of learning have the potential to be flexible, focused on increasing intrinsic motivation and centered on important, relevant life skills. So, the question begs to be asked: In light of recent research on best practices, has andragogy become a Band-Aid for poor pedagogy when teaching adults? Without inciting controversy, we believe that the need for a theory for adult learning is redundant, when current research on UDL addresses the six key characteristics of andragogy, while also capturing the most important aspects of pedagogy.

The UDL Framework encompasses the pedagogical shifts that may be necessary with adult learners without sacrificing the key characteristics of andragogy. The UDL Framework assumes variation in motivation, interest, and readiness, and so provides scaffolding in language function, explicit skills, and executive function, while keeping learner engagement and choice at the forefront. It is the ideal framework for all learners.

Knowles (1984) believed that flexibility was an essential part of andragogy. So, if we think of education as a continuum in which we teach children at one end of a spectrum and adults at the other, UDL is a bridge that is beneficial for K–12 learners and adult learners, because it requires high expectations, flexibility, and relevant, valuable learner participation. The following chart is our attempt to bring the two theories together, into one theory of learning, UDLagogy if you will.

@PoorRichardUDL · #UDLagogy is definitely a word. See that @NoahWebster?

TABLE A-1 UDL Brings Together Pedagogy and Andragogy

	PEDAGOGICAL	UNIVERSAL DESIGN FOR LEARNING	ANDRAGOGICAL
Concept of Learner	Responsibility of instructor	Learners experience autonomy, and are internally motivated to learn when they can realize their personal goals, values, and interests and see how they connect to learning. • Optimize individual choice and autonomy • Optimize relevance, value, and authenticity • Minimize threats and distractions	Student-directed with resistance to being "told"

TABLE A-1 UDL Brings Together Pedagogy and Andragogy *CONTINUED*

	PEDAGOGICAL	UNIVERSAL DESIGN FOR LEARNING	ANDRAGOGICAL
Role of Learner's Experience	Minimized—previous grade	Scaffold experiences for learners and provide options for learning. • Activate or supply background knowledge • Highlight patterns, critical features, big ideas, and relationships • Build fluencies with graduated levels of support for practice and performance • Guide information processing, visualization, and manipulation	Maximized (part of self-identity) and expected to be part of the process of learning new information
Readiness to Learn	Largely a function of age	Instructor provides scaffolding in executive function to prepare all learners • Guide appropriate goal setting • Support planning and strategy development • Enhance capacity for monitoring progress	Ready to learn based upon need to improve
Orientation to Learning	Subject-oriented and sequenced to logic of subject matter	Subject- and skill-orientation and relevance and engagement are at forefront • Heighten salience of goals and objectives • Optimize relevance, value, and authenticity	Life-, task-, or problem-oriented; sequenced to the logic of the life situation: "need to know," "relevant"

TABLE A-1 UDL Brings Together Pedagogy and Andragogy *CONTINUED*

	PEDAGOGICAL	UNIVERSAL DESIGN FOR LEARNING	ANDRAGOGICAL
Motivation to Learn	External	Progressively internal • Promote expectations and beliefs that optimize motivation • Facilitate personal coping skills and strategies • Vary demands and resources to optimize challenge • Foster collaboration and community • Increase mastery-oriented feedback	Internal
Format	Content plan	Optional paths provided for learners at all levels • Use multiple media for communication • Use multiple tools for construction and composition • Offer ways of customizing the display of information, alternatives for auditory information, and alternatives for visual information • Vary the methods for response and navigation • Optimize access to tools and assistive technologies	Process design

	PEDAGOGICAL	UNIVERSAL DESIGN FOR LEARNING	ANDRAGOGICAL
Instructor Role	"Sage-on-the-stage"	"Learning coach" • Foster collaboration and community • Increase mastery-oriented feedback • Maximize transfer and generalization	"Facilitator"
Learning Expectation	Predetermined by standard	Maximized for all learners • Heighten salience of goals and objectives • Optimize relevance, value, and authenticity	Potentially open-ended

Bringing this full circle, in UDL, the variability of learners is an integral part of the learning environment. If, as educators, we use the UDL Framework and hold the highest of expectations for all learners, K–12 children, and adults, we can engage, motivate, and ultimately facilitate authentic, relevant learning experiences for all learners.

It is this reason why we did not focus on the two theories of pedagogy and andragogy in this text. We simply discussed Universal Design for Learning. You, as the reader, can come to your own determination on the value of the conflicting theories, but given that UDL can bring the two together, we felt that the distinction was not worth examination.

UDL assumes learner variability, including age, and therefore, when implemented successfully, it will meet the needs of child and adult learners in a cloud environment.

References

Alsharif, N. Z., & Yongyue, Q. (2014). A three-year study of the impact of instructor attitude, enthusiasm, and teaching style on student learning in a medicinal chemistry course. *American Journal of Pharmaceutical Education, 78*(7), 1–13.

Archambault, L., Kennedy, K., & Bender, S. (2013). Cyber-truancy: Addressing issues of attendance in the digital age. *Journal of Research on Technology in Education, 46*(1), 1.

Arum, R., & Roksa, J. (2011). *Academically adrift: Limited learning on college campuses.* Chicago: University of Chicago Press.

Ashbaugh, M. L., (2013). Personal leadership in practice: A critical approach to instructional design innovation work. *TechTrends, 57*(5), 74–82.

Assor, A., Kaplan, H., & Roth, G. (2002). Choice is good, but relevance is excellent: Autonomy-enhancing and suppressing teacher behaviours predicting students' engagement in schoolwork. *British Journal of Educational Psychology, 72*(2), 261–278.

Bacon, D. R., & Stewart, K. A. (2006). How fast do students forget what they learn in consumer behavior? A longitudinal study. *Journal of Marketing Education, 28*(3), 181–192.

Baier, K., Hendricks, C., Warren Gorden, K., Hendricks, J. E., & Cochran, L. (2011). College students' textbook reading, or not! *American Reading Forum Annual Yearbook* [Online], Vol. 31.

Baker, C. (2010). The impact of instructor immediacy and presence for online student affective learning, cognition, and motivation. *Journal of Educators Online, 7*(1).

Barkley, J. M. (2006). Reading education: Is self-efficacy important? *Reading Improvement, 43*(4), 194–210.

Beatty, I. D. (2013). Standards-based grading in introductory university physics. *Journal of the Scholarship of Teaching and Learning, 13*(2), 1–22.

Beaudreau, S. A., & Finger, S. (2010). Critical Readings: Medical electricity and madness in the eighteenth century: The legacies of Benjamin Franklin and Jan Ingenhousz. In *Critical insights: Benjamin Franklin* (pp. 93–114). Pasadena, CA: Salem Press.

Blackmon, S. J. (2012). Outcomes of chat and discussion board use in online learning: A research synthesis. *Journal of Educators Online, 9*(2).

Blinderman, A. (1976). *Three early champions of education: Benjamin Franklin, Benjamin Rush, and Noah Webster.* Bloomington, IN: Phi Delta Kappa Educational Foundation.

Blinne, K. C. (2013). Start with the syllabus: HELPing learners learn through class content collaboration. *College Teaching, 61*(2), 41–43.

Bodie, G. D., Powers, W. G., & Fitch-Hauser, M. (2006). Chunking, priming and active learning: Toward an innovative and blended approach to teaching communication-related skills. *Interactive Learning Environments, 14*(2), 119–135.

Bork, R. H., & Rucks-Ahidiana, Z. (2013). Role ambiguity in online courses: An analysis of student and instructor expectations (CCRC Working Paper). New York, NY: Columbia University, Teachers College, Community College Research Center.

Callender, C. (2010). Is time an illusion? *Scientific American, 302*(6), 58-65.

Carr, N. (2012). The crisis in higher education. *MIT Technology Review, 115*(6), 32–40.

CAST (2011). Universal Design for Learning Guidelines version 2.0. Wakefield, MA: Author.

Chan, S. (2010). Applications of andragogy in multi-disciplined teaching and learning. *Journal of Adult Education, 39*(2), 25–35.

Chickering, A. W. & Gamson, Z. F. (1999). Development and adaptations of the seven principles for good practice in undergraduate education. *New Directions for Teaching and Learning, 1999*(80), 75.

Cleary, S. (2011). The ethos aquatic: Benjamin Franklin and the art of swimming. *Early American Literature, 46*(1), 51–67.

Common Core State Standards Initiative. (2015). *Key shifts in English language arts.* Retrieved from http://www.corestandards.org/other-resources/key-shifts-in-english-language-arts/

Daniel, F. & Braasch, J. L. G. (2013). Application exercises improve transfer of statistical knowledge in real-world situations. *Teaching of Psychology, 40*(3), 200–207.

Darling-Hammond, L., (2008). *Powerful learning: What we know about teaching for understanding.* San Francisco, CA: Jossey-Bass.

Davenport, J. (1987) Is there any way out of the andragogy mess? In M. Thorpe, R. Edwards, and A. Hanson (eds.), *Culture and processes of adult learning* (pp. 109–117). London: Routledge.

Dennen, V. P. (2007). Presence and positioning as components of online instructor persona. *Journal of Research on Technology in Education, 40*(1), 95–108.

Doabler, C. T., Clarke, B., Fien, H., Baker, S. K., Kosty, D. B., & Cary, M. S. (2015). The science behind curriculum development and evaluation: Taking a design science approach in the production of a tier 2 mathematics curriculum. *Learning Disability Quarterly, 38*(2), 97–111.

Educause. (2012). *7 things you should know about flipped classrooms.* Retrieved from https://net.educause.edu/ir/library/pdf/eli7081.pdf

Edyburn, D. L. (2010). Would you recognize universal design for learning if you saw it? Ten propositions for new directions for the second decade of UDL. *Learning Disability Quarterly, 33*(1), 33–41.

Fink, L. D. (2013). *Creating significant learning experiences: An integrated approach to designing college courses.* San Francisco, CA: Jossey-Bass.

Garrison, D. R., Anderson, T., & Archer, W. (2000). Critical inquiry in a text-based environment: Computer conferencing in higher education. *The Internet and Higher Education, 2*(2–3), 87–105.

Giacumo, L. A., Savenye, W., & Smith, N. (2013). Facilitation prompts and rubrics on higher-order thinking skill performance found in undergraduate asynchronous discussion boards. *British Journal of Educational Technology, 44*(5), 774–794.

Goodwin, B., & Miller, K. (2013). Evidence on flipped classrooms is still coming in. *Educational Leadership, 70*(6), 78.

Gruenbaum, E. A. (2010). Predictors of success for adult online learners: A review of the literature. *ELearn Magazine*. Retrieved from http://elearnmag.acm.org

Gumbrecht, J. (2015). Students, say goodbye to snow days—and say hello to school at home. *CNN*. Retrieved from http://www.cnn.com/2014/02/23/living/snow-days-virtual-schools/

Guo, P. (2013). Optimal video length for student engagement. *edX*. Retrieved from https://www.edx.org/blog/optimal-video-length-student-engagement

Hall, R. A. (2015). Critical thinking in online discussion boards: Transforming an anomaly. *Delta Kappa Gamma Bulletin, 81*(3), 21–27.

Hall, T.E., Strangman, N., & Meyer, A. (2003). Differentiated instruction and implications for UDL implementation. [Effective Classroom Practices Report.] Wakefield, MA: National Center on Accessing the General Curriculum.

Hawkins, A., Graham, C. R., Sudweeks, R. R., & Barbour, M. K. (2013). Academic performance, course completion rates, and student perception of the quality and frequency of interaction in a virtual high school. *Distance Education, 34*(1), 64–83.

Heerboth, M. K. & Mason, K. (2012). Educational materials can induce stereotype threat in elementary school students: Evidence for impaired math performance following exposure to a token. *Psychology Journal, 9*(4), 120–128.

Hilferty, F., Redmond, G., & Katz, I. (2010). The implications of poverty on children's readiness to learn. *Australasian Journal of Early Childhood, 35*(4), 63–71.

Hobson, E. H. (2004). *IDEA Paper #40: Getting students to read: Fourteen tips.* Retrieved from http://www.idea.ksu.edu/

Hodges, C. B., & Cowan, S. F. (2012). Preservice teachers' views of instructor presence in online courses. *Journal of Digital Learning in Teacher Education, 28*(4), 139–145.

Irvin, M. J., Hannum, W. H., de la Varre, C., & Farmer T. W. (2010). Barriers to distance education in rural schools. *Quarterly Review of Distance Education, 11*(2), 73–90.

Jaggars, S. S., Edgecombe, N., & Stacey, G. W. (2013). *Creating an effective online instructor presence.* New York: Community College Research Center, Columbia University.

Jaggars, S. S. (2012). Online learning in community colleges. In M. G. Moore (ed.), *Handbook of distance education.* 3rd ed. (pp. 594–608). New York: Routledge.

Ke, F. (2010). Examining online teaching, cognitive, and social presence for adult students. *Computers & Education, 55*(2), 808–820.

Keramidas, C. G. (2012). Are undergraduate students ready for online learning? A comparison of online and face-to-face sections of a course. *Rural Special Education Quarterly, 31*(4), 25–32.

Kerka, S. (1994). *Self-directed learning: Myths and realities* (ERIC Document Reproduction Service No. ED 365 818). Washington, DC: Office of Educational Research and Improvement.

Knowles, M. (1980). *The modern practice of adult education: From pedagogy to andragogy.* Englewoods Cliffs, NJ: Cambridge Adult Education.

Knowles, M. (1984). *Andragogy in Action.* San Francisco, CA: Jossey-Bass.

Knowles, M. (1989). *The making of an adult educator: An autobiographical journey* (1st ed.). San Francisco, CA: Jossey-Bass.

Knowles, M., Holton, E. F., & Swanson, R. A. (1998). *The adult learner: The definitive classic in adult education and human resource development* (5th ed.). Houston, TX: Gulf Publishing Company.

Krell, E. (2010). A head start on real world applications. *Baylor Business Review, 29*(1), 6–13.

Lawson, K. G. (2005). Using eclectic digital resources to enhance instructional methods for adult learners. *OCLC Systems and Services: International digital library perspectives, 21*(1), 49–60.

Lowe, G., & Belcher, S. (2012). Direct instruction and music literacy: One approach to augmenting the diminishing? *Australian Journal of Music Education, 1,* 3–13.

Mayne, L. A., & Wu, Q. (2011). Creating and measuring social presence in online graduate nursing courses. *Nursing Education Perspectives, 32*(2), 110–114.

Mazzolini, M. & Maddison, S. (2007). When to jump in: The role of the instructor in online discussion forums. *Computers & Education, 49*(2), 193–213.

McGrath, V. (2009). Reviewing the evidence on how adult students learn: An examination of Knowles' model of andragogy. *Adult Learner: The Irish Journal of Adult and Community Education,* 99–110.

McWilliam, E. (2008). Unlearning how to teach. *Innovations in Education and Teaching International, 45*(3), 263–269.

Meyer, A., Rose, D. H., & Gordon, D. (2014). *Universal design for learning: Theory and practice.* Wakefield, MA: CAST Professional Publishing.

National Center for Education Statistics (NCES). (2014). *Web tables: Enrollment in distance education courses, by state: Fall 2012* (NCES no. 2014023). Retrieved from http://nces.ed.gov/pubs2014/2014023.pdf

National Telecommunications & Information Administration (NTIA). (2015). *National Broadband Map*. Retrieved from http://www.ntia.doc.gov/

Novak, K. (2014). Paideia Seminars: How to build student collaboration skills. *The Teaching Channel*. Retrieved from https://www.teachingchannel.org/blog/2014/05/21/paideia-seminars-how-to-build-student-collaboration-skills/

Ocak, G., & Yamac, A. (2013). Examination of the relationships between fifth graders' self-regulated learning strategies, motivational beliefs, attitudes, and achievement. *Educational Sciences: Theory and Practice, 13*(1), 380–387.

Orr, R., Williams, M. R., & Pennington, K. (2009). Institutional efforts to support faculty in online teaching. *Innovative Higher Education, 34*(4), 257–268.

Orso, D., & Doolittle, J. (2012). Instructor characteristics that affect online student success. *Faculty Focus*. Retrieved from http://www.facultyfocus.com/articles/online-education/instructor-characteristics-that-affect-online-student-success/

Palenque, S. M., & DeCosta, M. (2014). The art and science of successful online discussions. *Faculty Focus*. Retrieved from http://www.facultyfocus.com/articles/online-education/art-science-successful-online-discussions/

Park, J-H., & Choi, H. J. (2009). Factors influencing adult learners' decision to drop out or persist in online learning. *Educational Technology & Society, 12*(4), 207–217.

Pfister, H-R., & Oehl, M. (2009). The impact of goal focus, task type and group size on synchronous net-based collaborative learning discourses. *Journal of Computer Assisted Learning, 25*(2), 161–176.

Reupert, A., Maybery, D., Patrick, K., & Chittleborough, P. (2009). The importance of being human: Instructors' personal presence in distance programs. *International Journal of Teaching and Learning in Higher Education, 21*(1), 47–56.

Roper, A. R. (2007). How students develop online learning skills. *Educause Review*. Retrieved from http://www.educause.edu/ero/article/how-students-develop-online-learning-skills

Roscorla, T. (2014, November 4). 6 Tips for online learning days in stormy weather. *Digital Education*. Retrieved from http://www.centerdigitaled.com/news/6-Tips-for-Online-Learning-Days-in-Stormy-Weather.html

Rose, D. H., Harbour, W. S., Johnston, C. S., Daley, S. G., & Abarbanell, L. (2006). Universal design for learning in postsecondary education: Reflections on principles and their application. *Journal of Postsecondary Education and Disability, 19*(2), 17.

Rose, D. H., & Meyer, A. (2006). Preface. In D. H. Rose & A. Meyer (eds.), *A practical reader in universal design for learning* (pp. vii–xi). Cambridge, MA: Harvard Educational Press.

Ross, J. A. (1995). Strategies for enhancing teachers' beliefs in their effectiveness: Research on a school improvement hypothesis. *Teachers College Record, 97*(2), 227–251.

Rourke, L., Anderson, T., Garrison, D. R., & Archer, W. (2001). Assessing social presence in asynchronous text-based computer conferencing. *Journal of Distance Education, 14*(2), 51–70.

Sanacore, J. (2008). Turning reluctant learners into inspired learners. *Clearing House: A Journal of Educational Strategies, 82*(1), 40–44.

Schelly, C. L., Davies, P. L., & Spooner, C. L. (2011). Student perceptions of faculty implementation of Universal Design for Learning. *Journal of Postsecondary Education and Disability, 24*(1), 17–30.

Shachar M., & Neumann, Y. (2010). Twenty years of research on the academic performance differences between traditional and distance learning: Summative meta-analysis and trend examination. *MERLOT Journal of Online Learning and Teaching, 6*(2).

Sharpe, S. (2005). Why online teaching turned me off. *The Washington Post.* pW31. Retrieved from http://www.washingtonpost.com/wp-dyn/articles/A11196-2005Mar29.html

Silva, E., White, T., & Toch, T. (2015). *The Carnegie unit: The century-old standard in a changing education landscape.* Stanford, CA: Carnegie Foundation for the Advancement of Teaching.

Sobel, D. (1995). Mihaly Csikszentmihalyi. *Omni, 17*(4), 73.

Tata, J. (1999). Grade distributions, grading procedures, and students' evaluations of instructors: A justice perspective. *Journal of Psychology, 133*(3), 263.

Taylor, B., & Kroth, M. (2009). Andragogy's transition into the future: Meta-analysis of andragogy and its search for a measurable instrument. *Journal of Adult Education, 38*(1), 1–11.

Thompson, G. (2014). 1-to-1 + BYOD + PD = Success. *T H E Journal, 41*(8), 13.

Tu, Y., & Soman, D. (2014). The categorization of time and its impact on task initiation. *Journal of Consumer Research, 41*(3), 810–822.

Vasser, N. (2010). Instructional design processes and traditional colleges. *Online Journal of Distance Learning Administration, 13*(4). Retrieved online at http://www.westga.edu/~distance/ojdla/winter134/vasser134.html

Vella, F. (1990). The changing face of college teaching. [Review of the journal article "Changing the face of your teaching" by M.D. Svinicki] *New Directions for Teaching and Learning, 1990*(42), 135.

Vygotsky, L. S. (1978). *Mind in society: The development of higher psychological processes.* Cambridge, MA: Harvard University Press.

Westermann, E. B. (2014). A half-flipped classroom or an alternative approach? Primary sources and blended learning. *Educational Research Quarterly, 38*(2), 43–57.

Wise, A., Chang, J., Duffy, T., & Del Valle, R. (2004). The effects of teacher social presence on student satisfaction, engagement, and learning. *Journal of Educational Computing Research, 31*(3), 247–271.

Xu, D. & Jaggars, S. S. (2014). Performance gaps between online and face-to-face courses: Differences across types of students and academic subject areas. *The Journal of Higher Education, 85*(5), 633–659.

Zafft, C., Kallenbach, S., & Spohn, J. (2006). *Transitioning adults to college: Adult basic education program models.* Cambridge, MA: National Center for the Study of Adult Learning and Literacy.

Index

self-regulation, 145
symbols, 147
June. *See also* students
  barriers, 13, 15
  comprehension, 144
  effort and persistence, 142
  engagement, 32
  executive functions, 143
  expression and communication, 143
  facilitating discourse with, 81
  impact of syllabus revisions, 68
  language, 144
  mathematical expressions, 144
  perception, 144
  physical action, 143
  providing options for, 96
  recruiting interest, 142
  self-regulation, 142
  symbols, 144

K

Ke, F., 75
Keramidas, C. G., 6, 20
Khan Academy videos, 124
Knowles, Holton, & Swanson, 156–157
Knowles, M., 21, 155–156
  andragogy versus pedagogy, 157–158
  concept of learner in pedagogy, 159
  flexibility of andragogy, 162
  format of lessons, 160
  learner's experience, 159
  learning expectations, 161
  motivation to learn, 160
  orientation to learning, 160
  readiness of adults to learn, 159–160
  role of instructor, 160–161
Krell, E., 93
Kriti. *See also* students
  barriers, 13, 15
  comprehension, 140
  effort and persistence, 138
  executive functions, 139

expression and communication, 139
impact of syllabus revisions, 68
language, 140
mathematical expressions, 140
perception, 141
physical action, 140
providing options for, 96
recruiting interest, 138
self-regulation, 138
symbols, 140

L

Lagace, Martha, 65–66
language
  options for Coco, 136
  options for José, 147
  options for June, 144
  options for Kriti, 140
  options for Ray, 151
Lawson, K. G., 10
leadership, importance in instructional design, 80
learners. *See also* students
  purposeful type of, 22–23
  resourceful type of, 22–23
  self-direction of, 9
  self-regulation of, 10
  strategic type of, 22–23
  variability among, 123
learning, pedagogy versus andragogy, 21
learning needs, managing, 32
learning options, supporting and engaging, 90
lectures, recording, 124
lessons, planning, 63
LMSes (learning management systems)
  examples of, 4
  including personal information in, 43–44
  and virtual snow days, 125–130
Lowe & Belcher, 82

M

About the Authors

Katie Novak is the assistant superintendent of the Groton-Dunstable Regional School District in Massachusetts and one of the country's leading experts in Universal Design for Learning (UDL). With 13 years of experience in teaching and administration and an earned doctorate in curriculum and teaching, Novak designs and presents workshops both nationally and internationally, focusing on implementation of UDL and the Common Core. She is the author of *UDL Now! A Teacher's Monday-Morning Guide to Implementing Common Core Standards Using Universal Design for Learning* (CAST Professional Publishing, 2014).

Tom Thibodeau is an assistant provost at New England Institute of Technology (NEIT) and former president of the New England Faculty Development Consortium. A video editor and videographer by trade, Tom transitioned to education over 25 years ago. In 1997, as an education technologist at NEIT, he began developing online courses to meet the needs of a wide variety of learners. With close to 20 years of experience in online education and faculty development, Tom's expertise in the design and delivery of accessible, inclusive virtual courses makes him a leader in the field.

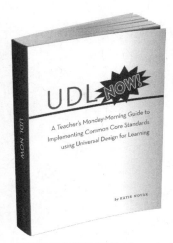